PROCEEDINGS

of the

NATURAL

PHILOSOPHY

ALLIANCE

Volume 11

Printed in the United States of America.
ISBN 978-1517770877

A Letter from the Editor

The Proceedings is a collection of papers from NPA members that were presented at the 21st Annual NPA Conference held November 19-21, 2014 in Baltimore, MD USA. NPA Conferences attract scientists from around the world, who work in many different areas of science, and are interested in pursuing alternatives to the mainstream. The present proceedings include papers on critical analysis of the theory of relativity, gravitational and climate anomalies, alternative explanations of some astronomical phenomena, and the causal interpretation of quantum mechanics. Some of the ideas presented here might appear to be incorrect, while on the other hand, many could develop into new areas of physics in the future. NPA authors follow uncharted paths, and we cannot require the same scientific accuracy from research publications on new concepts as we expect from works in mainstream science. Publishing works with new concepts is absolutely necessary in order to chart new directions and to avoid monopolies in physics.

Nina Sotina

Editor

Table of Contents

Review of Experiments that Contradict Special Relativity and Support neo-Lorentz Relativity: Latest Technique to Detect Dynamical Space Using Quantum Detectors

Reginald T. Cahill

School of Chemical and Physical Sciences, Flinders University, Adelaide 5001, Australia

The anisotropy of the velocity of EM radiation has been repeatedly detected, including the Michelson-Morley experiment of 1887, using a variety of techniques. The experiments reveal the existence of a dynamical space that has a velocity of some 500km/s from a southerly direction. These consistent experiments contradict the assumptions of Special Relativity, but are consistent with the assumptions of neo-Lorentz Relativity. The existence of the dynamical space has been missed by physics since its beginnings. Novel and checkable phenomena then follow from including this space in Quantum Theory, EM Theory, Cosmology, etc, including the derivation of a more general theory of gravity as a quantum wave refraction effect. The corrected Schrödinger equation has resulted in a very simple and robust quantum detector, which easily measures the speed and direction of the dynamical space. This report reviews the key experimental evidence.

1. Introduction

Determining the nature of reality, in the sense of discovering the properties of space, time, matter, has been a longstanding problem since the time of the ancient Greek physicists, who argued, theoretically, long and slower clock, light speed is either faster or space-time is relatively greater. We assume the latter. Consider also gravitational. At that time a majority of scientists did not introduced geometrical models for space and time, which were adopted by Newton, and which persisted until the most famous of all experiments by Michelson and Morley in 1887. That experiment was conceived as a technique to determine the velocity of the earth through the aether, a substance that supposedly filled Galileo's geometrical space. Almost all physics publications assert that this interferometer experiment produced a null result, namely that on rotation the fringe shifts did not move, which would have been the case if an aether was flowing past the device, and for which the speed of light was fixed relative to the aether. This putative null result was used by Einstein in 1905 to initiate a new theory for space and time, namely that space and time were not separate phenomena, but were united into space-time, in which there is no observer independent notion of space or time, and no aether. This space-time model has persisted until the present time, and has completely determined the understanding of reality by academic physicists. However the Michelson-Morley 1887 paper [33] reveals observed fringe shifts and a speed of up to 10km/s. which was less than expected. But the key point is that they used Galilean-Newtonian physics to calibrate the interferometer, which provided the computation from the observed fringe shifts to a speed of translation of the earth. So the data was indicating the failure of that calibration theory, and not that the results amounted to a null effect. Another mistake by MM was to average data from different days, but at the same local solar time. So MM missed the discovery also of gravitational waves, for which the averaging washed out the wave effects. However in 2002 Cahill and Kitto [5,7] used Lorentz relativity to recalibrate the sensitivity of the MM detector. It was then discovered that the Michelson interferometer had a fundamental design flaw, that had gone unnoticed since 1887, namely that the device had zero sensitivity unless operated with a dielectric present in the light paths. Indeed the 1887 experiment had air present, and the new calibration theory implied that it was some 2000 times less sensitive than assumed by

MM. Some of MM 1887 data is shown in Fig.3, together with some data from the 1925/26 Miller [34] interferometer experiment. So the starting point for Einstein's space-time model, whose key assumption is that the speed of light is the same in all directions and for all observers, is contradicted by the earliest experiment. All Michelson interferometer experiments in recent years use vacuum mode, and so have zero sensitivity to light speed anisotropy. So some 100 years after Einstein's innovation of 'space-time', it has been necessary to review all relevant experiments, to develop new experimental techniques, and to rebuild the foundations of physics. The new foundations involve a dynamical space, with no aether, and this theory has been tested against data from numerous experiments and earth and astronomical observations. A key development has been the discovery of a new theory of gravity, namely that it is caused by quantum matter waves being refracted by inhomogeneities and time dependencies of the velocity field that describes the dynamical space. Until recently all experiments to detect and characterize the dynamical space have used either light speed or RF EM wave speed anisotropies. However in 2013 a quantum effect was discovered, in which the passing dynamical space modifies the current through a reverse-biased Zener diode, Cahill [23]. This has made the determination of the flow of space essentially very cheap, simple and robust. The observed fluctuations are actual gravitational waves, but not with the characteristics predicted by General Relativity, which have never been detected. All the non-null experimental data from 1887 to 2014 now agree wrt the speed and direction of the earth through the dynamical space, and only some of that data is reviewed here. As well new phenomena caused by the fluctuations in the flow of space are now being discovered. The most significant is that the dynamical space does not have a measure of energy, but can induce energy in matter systems. This means that the Conservation of Energy Principle, namely the 1st Law of Thermodynamics, does not apply to space-flow turbulence /gravitational-wave induced effects [27,28].

2. Relativity Theories

A "Relativity Principle" (RP) specifies how observations by different observers are related. In doing so the RP reflects fundamental aspects of realty, and any proposed RP is subject to ongoing experimental challenge.

There have been three major relativity theories: Galileo Relativity (GaR), Lorentz Relativity (LR) and Einstein Special Relativity (SR), with the later much celebrated, while the LR is essentially ignored. Yet it is often incorrectly claimed that LR and SR are experimentally indistinguishable. It has been shown [13,22] that (i) they are experimentally distinguishable, (ii) that comparison of gas-mode Michelson interferometer experiments with spacecraft earth-flyby Doppler shift data [22] demonstrate that it is LR that is consistent with the data, while SR is in conflict with the same data, (iii) SR is exactly derivable from Galilean Relativity by means of change of space and time coordinates, so that the well-known SR relativistic effects are purely coordinate effects, and cannot correspond to the observed dynamical relativistic effects. The connections between these three relativity theories has become apparent following the discovery that space is a dynamical and observable system, and that space and time are distinct phenomena.

We give a non-historical presentation, because historical presentations were always confused by the lack of realization that a dynamical space existed, although serious consideration was given to Lorentz Relativity [2-4].

But 1st a warning: a common error when discussing the physics of space and time is to confuse space and time coordinates with the actual phenomenon of space and time, and also to confuse space intervals, as measured by a ruler or round trip light speed measurements, and time measured by an actual clock, with actual intrinsic measures of space and time phenomena: coordinates are arbitrary, whereas the intrinsic measures are set by the dynamics of space.

3. Galilean Relativity

We give here a modern statement of Galilean Relativity. The assumptions in GaR are (i) space exists, but is not observable and not dynamical, and is modelled as a Euclidean 3-space (E^3), which entails the notion that space is without structure, (ii) observers measure space and time intervals using rods and clocks, whose respective lengths and time intervals are not affected by their motion through space, (iii) the speed of light (in vacuum) is fixed at c wrt the space, and (iv) velocities are measured relative to observers, where different observers, O and O', relate their space and time coordinates by

$$t' = t, \quad x' = x - Vt, \quad y' = y, \quad z' = z \qquad (1)$$

where V is the relative speed of the observers (in their common x-direction, for simplicity). The speed w of an object or waveform (in the x direction) according to each observer, is related by

$$w' = w - V \qquad (2)$$

Eqs (1) and (2) form the Galilean Relativity Transformation, and the underlying assumptions define Galilean Relativity (GaR). Newton based his dynamics on Galilean Relativity, in particular his theory of gravity, to which General Relativity reduces in the limit of low mass densities and low speeds.

4. Lorentz and Neo-Lorentz Relativity

When Maxwell formulated his unification of electric and magnetic fields[*] the speed of EM waves came out to be the constant $c = 1/\sqrt{\varepsilon_0\mu_0}$ for any observer, and so independent of the motion of the observers wrt one another or to space. This overtly contradicted GaR, in (2). Hertz in 1890 [29] pointed out the obvious fix-up, namely that Maxwell had mistakenly not used the then-known Euler constituent derivative $\frac{\partial}{\partial t} + \mathbf{v} \cdot \Delta$ in place of $\frac{\partial}{\partial t}$ where $\mathbf{v}$ is the velocity of some structure to space relative to an observer, in which case Maxwell's equations would only be valid in the local rest frame defined by this structure. In that era a dual model was then considered, namely with a Euclidean space E^3 and an extended all-filling aether substance, so that the velocity $\mathbf{v}$ was the velocity of the aether relative to an observer. To be explicit let us consider the case of electromagnetic waves, as described by the vector potential $\mathbf{A}(\mathbf{r},t)$ satisfying the wave equation (in absence of charges and currents), but using the Euler constituent derivative, as suggested by Hertz:

$$\left(\frac{\partial}{\partial t} + \mathbf{v}(\mathbf{r},t)\nabla\right)^2 \mathbf{A}(\mathbf{r},t) = c^2\nabla^2\mathbf{A}(\mathbf{r},t). \qquad (3)$$

Here $\nabla = \{\frac{\partial}{\partial x}, \frac{\partial}{\partial y}, \frac{\partial}{\partial z}\}$. In Lorentz Relativity there is *an* aether in addition to an actual Euclidean space, and $\mathbf{v}$ is independent of $\mathbf{r}$ and t; whereas in neo-Lorentz Relativity $\mathbf{v}(\mathbf{r},t)$ describes a dynamical space, with $\mathbf{r}$ and t describing a coordinate system for space and time. We find plane-wave solutions only for the case where the

space flow velocity, relative to an observer, is locally time and space independent, *viz* uniform,

$$\mathbf{A}(\mathbf{r},t) = \mathbf{A}_0 \sin(\mathbf{k}\cdot\mathbf{r} - \omega t)$$

with $\omega(\mathbf{k},\mathbf{v}) = c|\mathbf{k}| + \mathbf{v}\cdot\mathbf{k}$. The EM wave group velocity is then

$$\mathbf{v}_g = \nabla_k \omega(\mathbf{k},\mathbf{v}) = c\hat{\mathbf{k}} + \mathbf{v}$$

and we see that the wave has velocity $\mathbf{v}_g$ relative to the observer, with the space flowing at velocity $\mathbf{v}$ also relative to the observer, and so the EM speed is c in direction $\hat{\mathbf{k}}$ relative to the aether (LR) or dynamical space (nLR). In searching for experimental evidence for the existence of this aether, or more generally a Preferred Frame of Reference (PFR), Michelson conceived of his interferometer [33]. Unknown to Michelson was that his design had an intrinsic fatal flaw: if operated in vacuum mode it was incapable of detecting the PFR effect, while with air present, as operated by Michelson and Morley in 1887, it was extremely insensitive [7,8]. The problem was that Michelson had used Newtonian physics, *viz* GaR, in calibrating the interferometer. Michelson and Morley detected fringe shifts, but they were smaller than expected, and were interpreted as a null effect: there was no aether or PFR effect. However Lorentz [30,31] and Fitzgerald [32] offered an alternative explanation: physical objects, such as the arms supporting the interferometer optical elements, undergo a contraction in the direction of movement through the aether, or more generally relative to the PFR: the length becoming $L = L_0\sqrt{1 - v_R^2/c^2}$, where L_0 is the physical length when at rest wrt the PFR, and v_R is the speed relative to the PFR. It must be noted that this is not the Lorentz contraction effect predicted by SR, as discussed later, as that involves $L = L_0\sqrt{1 - v_O^2/c^2}$, where v_O is the speed of the arm or *any* space interval relative to the observer. The difference between these two predictions is stark, and has been observed experimentally, and the SR prediction is proven wrong.

Next consider two observers, O and O', in relative motion. Then the actual intrinsic or physical time and space coordinates of each are, in both LR and nLR, related by the Galilean transformation, and here we consider only a uniform $\mathbf{v}$: these coordinates are not the directly measured distances/time intervals - they require corrections to give the intrinsic values. We have taken the simplest case where V is the intrinsic relative speed of the two observers in their common x directions. Then from (1) the derivatives are related by

---

[*]The now standard formalism was actually done by Heaviside.

$$\frac{\partial}{\partial t}=\frac{\partial}{\partial t'}-V\frac{\partial}{\partial x'}, \ \frac{\partial}{\partial x}=\frac{\partial}{\partial x'}, \ \frac{\partial}{\partial y}=\frac{\partial}{\partial y'}, \ \frac{\partial}{\partial z}=\frac{\partial}{\partial z'}.$$

In the general case space rotations may be made. Then (3) becomes for the 2nd observer, with $v'=v-V$,

$$\left(\frac{\partial}{\partial t'}+\mathbf{v}'\cdot\nabla'\right)^2\mathbf{A}'(\mathbf{r}',t')=c^2\nabla'^2\mathbf{A}'(\mathbf{r}',t'). \quad (4)$$

with $\mathbf{A}'(\mathbf{r}',t')=\mathbf{A}(\mathbf{r},t)$. If the flow velocity $\mathbf{v}(\mathbf{r},t)$ is not uniform then we obtain refraction effects for the EM waves, capable of producing gravitational lensing. Only for an observer at rest in a time independent and uniform aether (LR) or dynamical space (nLR) does v' disappear from (4).

5. Special Relativity from Galilean Relativity

The above uses physically intrinsic choices for the time and space coordinates, which are experimentally accessible. However we could choose to use a new class of time and space coordinates, indicated by upper-case symbols T, X, Y, Z, that mixes the above time and space coordinates. We begin by showing that Special Relativity (SR), with its putative spacetime as the foundation of reality, is nothing more than Galilean Relativity (GaR) written in terms of these mixed space and time coordinates. The failure to discover this, until 2008 [13] reveals one of the most fundamental blunders in physics. One class of such mixed coordinates for O is[†]

$$T=\gamma(v)\left((1-\frac{v^2}{c^2})t+\frac{vx}{c^2}\right),$$
$$X=\gamma(v)x, \ Y=y, \ Z=z \quad (5)$$

where $\mathbf{v}$ is the uniform speed of space (in the x direction), and where $\gamma(v)=1/\sqrt{1-v^2/c^2}$. Note that this is not a Lorentz transformation. If an object has speed w, $x=wt$, wrt to O, then it has speed W, $X=WT$, using the mixed coordinates, wrt O

$$W=\frac{w}{1-\frac{v^2}{c^2}+\frac{v}{c^2}w} \quad (6)$$

Similarly for O' using v', w' and W'. In particular (6) gives for the relative speed of O' wrt O in the mixed coordinates

$$\overline{V}=\frac{V}{1-\frac{V^2}{c^2}+\frac{v}{c^2}V} \quad (7)$$

Using the above we may now express the Galilean speed transformation (2) in terms of W', W and $\overline{V}$ for the mixed coordinates, giving

$$W'=\frac{W-\overline{V}}{1-W\overline{V}/c^2} \quad (8)$$

which is the usual SR transformation for speeds, but here derived exactly from the Galilean transformation. Note that c enters here purely because of the definitions in (5), which is designed to ensure that wrt the mixed space-time coordinates the speed of light is invariant: c. To see this note that from ((5) the transformations for the derivatives are found to be

$$\frac{\partial}{\partial t}=\gamma(v)\left(1-\frac{v^2}{c^2}\right)\frac{\partial}{\partial T}$$
$$\frac{\partial}{\partial x}=\gamma(v)\left(\frac{v}{c^2}\frac{\partial}{\partial T}+\frac{\partial}{\partial X}\right)$$
$$\frac{\partial}{\partial y}=\frac{\partial}{\partial Y}, \ \frac{\partial}{\partial z}=\frac{\partial}{\partial Z}. \quad (9)$$

$\overline{\nabla}=\{\frac{\partial}{\partial X},\frac{\partial}{\partial Y},\frac{\partial}{\partial Z}\}$. Then we have from (3) for uniform $\mathbf{v}$,

$$\left(\frac{\partial}{\partial T}\right)^2\overline{\mathbf{A}}(\mathbf{R},T)=c^2\overline{\nabla}^2\overline{\mathbf{A}}(\mathbf{R},T).$$

with $\mathbf{R}=\{X,Y,Z\}$ and $\overline{\mathbf{A}}(\mathbf{R},T)=\mathbf{A}(\mathbf{r},t)$.

The speed of EM waves is now c for all observers. This is a remarkable result. In the new class of coordinates the dynamical equation no longer contains the space velocity $\mathbf{v}$ - it has been mapped out of the dynamics. The EM dynamics is now invariant under Lorentz transformations.

$$T'=\gamma(\overline{V})\left(T-\frac{\overline{V}X}{c^2}\right),$$
$$X'=\gamma(\overline{V})(X-\overline{V}T), \ Y'=Y, \ Z'=Z, \quad (10)$$

and we note that for two events with coordinate differences $\{dT, dX\}$ or $\{dT', dX'\}$

$$dl^2 \equiv c^2 dT'^2 - dX'^2 = c^2 dT^2 - dX^2 \qquad (11)$$

defines the invariant interval for different observers.

Figure 1: Here is derivation of SR length contraction from Galilean Relativity using coordinates introduced in ((5). Consider two events: (1) RH end of rod travelling with observer O, with speed W wrt observer O', passes O', and (2) when LH end passes O'. Then $dX'=0$, and $L'=WdT'$ defines L'. For O $dX=L$ and $L=WdT$. Then (11) gives, $L' = L\sqrt{1 - \frac{w^2}{c^2}}$ with W the speed of the rod wrt O'. However this is purely a coordinate effect, and has no physical significance. Experiment shows that it is the speed of the rod wrt space, v_R, that actually determines the length contraction

There is now no reference to the underlying flowing space: for an observer using this class of space and time coordinates the speed of EM waves relative to the observer is always c and so invariant - there will be no EM speed anisotropy. We could also introduce, following Minkowski, "spacetime" light cones along which $d\tau^2 = dT^2 - d\mathbf{R}^2/c^2 = 0$. Note that $d\tau^2$ is invariant under the Lorentz transformation (10). Then pairs of spacetime events could be classified into either time-like $d\tau^2 > 0$, or space-like, $d\tau^2 \leq 0$, with the time ordering of spacelike events not being uniquely defined. However this outcome is merely an artefact of the mixed space-time coordinates: dT is not the actual time interval.

Confusing a space and time coordinate system with actual space and time phenomena has confounded physics for more than 100 years, with this illustrated above by the recently discovered exact relationship between Galilean Relativity and Einstein Relativity. In mainstream physics it is claimed that Special Relativity reduces to Galilean Relativity only in the limit of speeds small compared to c. But the various so-called "relativistic effects" ascribed to Special Relativity are nothing more than coordinate effects - they are not real. It was Lorentz who first gave a possible *dynamical* account of relativistic effects, namely that they are caused by absolute motion of objects relative to the aether (LR) or, now, dynamical space (nLR), which according to the evidence discussed above, is absolute motion relative to a dynamical and structured quantum

foam substratum: space. In Lorentz Relativity relativistic effects are genuine dynamical effects and must be derived from some dynamical theory. This has yet to be done, and for the length contraction effect would involve the quantum theory of matter.

Finally we note in Fig.1 that the so-called length contraction effect in SR is exactly derivable from GaR - and so it is purely a coordinate effect, and so has no physical meaning.

We note that the Lorentz transformation (10) is not relevant to nLR.

6. Detecting Lorentz Relativistic Effects

We now show how only Lorentz Relativity gives a valid account of the experimental results dealing with light speed anisotropy. To that end we consider the differing predictions made by the relativity theories for the length contraction effect, and we use data from Michelson interferometer experiments, which being a 2nd order in v/c detector requires length contraction effects to be included, when relevant. These contradictory predictions are compared with detailed data from the NASA spacecraft earth-flyby Doppler shifts, which in LR and nLR do not involve any length contraction, as no objects/supporting arms are involved. The flyby Doppler shifts have been also confirmed by laboratory 1st order v/c experiments by DeWitte [12] and Cahill [20], and so not requiring 2nd order v/c length contraction effects to be considered. So we have a critical and decisive test of the relativity theories. In all cases we parametrize the calibration theory for the Michelson interferometer travel time difference between the two arms according to

$$\Delta t = k^2 \frac{L_0 v_p^2}{c^3} \cos(2\theta) \qquad (12)$$

where k^2 is the theory-dependent calibration constant. Here L_0 is the at-rest arm length, v_P is the relevant velocity projected onto the plane of the interferometer, and θ is the angle between that projected velocity and one of the arms, see Fig.2.

Lorentz and neo-Lorentz Relativity Interferometer Calibration. In both LR and nLR the length contraction effect is a real dynamical effect caused by the absolute motion of an actual object wrt aether (LR) or dynamical space (nLR). A simple analysis yields the calibration constant $k^2 = (n^2 - 1)$, when $n \approx 1$ is the refractive index of the gas present: for air $n=1.00029$ at STP, giving $k^2=0.00058$. Some data from the Michelson-Morley and Miller experiments are shown in Fig.3, showing, together with other data, that this value of k^2 gives excellent agreement with the Doppler shift data, and different 1st order in v/c experiments [20]. The gas-mode interferometer experiments and spacecraft Doppler shift data give $v\approx500$km/s. Note that high-accuracy vacuum-mode Michelson interferometers will give a null result ($n=1$), as has been repeatedly observed.

Galilean Relativity Interferometer Calibration. In Galilean Relativity there is no length contraction effect, and repeating the analysis, without that effect, we obtain $k^2=n^3$ (≈1 for air). This is the calibration constant used by Michelson-Morley in 1887. Using this to analyze their data they found that $v_P\leq10$km/s. This is in stark conflict with the speed of $v\approx500$km/s from spacecraft earth-flyby Doppler shift and 1st order in v/c experiments. So Galilean Relativity is ruled out.

Einstein Relativity Interferometer Calibration. There are two routes to k^2 from Einstein Relativity, depending on which choice of space and time variables is used. Here we use the Galilean space and time coordinates, as we have shown that they are the physical coordinates that underly SR, in which case $k^2=n^3$, giving $v_P\leq10$ km/s and so again is in stark disagreement with experimental data.

In a different approach we use the mixed space and time coordinates conventionally used in SR calculations. Then the speed of light is c/n - invariant wrt to these coordinates, but there is no length contraction effect, because the arms are at rest wrt the observer. Then again we find that $k^2=n^3\approx1$, and in disagreement with the experimental data.

7. Michelson Interferometer Detectors

The Michelson interferometer was a brilliantly conceived instrument for measuring light speed anisotropy. However Michelson made two critically incorrect assumptions, which inadvertently had the effect of misguiding physics for another 100 years and more. The 1st was to assume Newtonian physics in determining the calibration theory for the instrument, and the 2nd was to average data from successive days at the same approximate times, with the assumption being that this would average out "instrumental fluctuations", when it had the opposite effect because there were significant "gravitational wave" effects in the data, and these were different on different days, even at the same time.

We now have a clear understanding of the design principles of the Michelson interferometer as a detector of light speed anisotropy, and *ipso facto* as a detector for the actual 3-space flow turbulence/ gravitational waves, (Cahill and Kitto, [5]), (Cahill, [6,7]). This is because two different and independent effects exactly cancel in vacuum mode. The key insight is that the dynamical space is describable at a macroscopic/classical level by a detectable velocity field $\mathbf{v}(\mathbf{r},t)$, relative to an observer using spatial coordinate $\mathbf{r}$ and time coordinate t, both of which must be carefully determined so as to remove absolute motion effects, that is, effects caused by the motion of rods and clocks wrt space. The key aspects of the interferometer are shown in Fig.1. Taking account of the geometrical path differences, the Fitzgerald-Lorentz arm-length contraction and the Fresnel drag effect leads to the travel time difference between the two arms, and which is detected by interference effects[‡], is given by

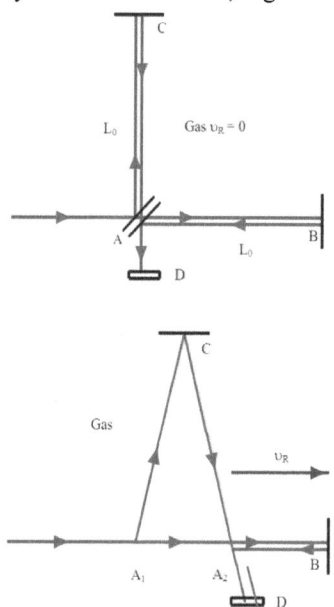

[‡]The dielectric of course does not cause the observed effect, it is merely a necessary part of the instrument design physics, just as mercury in a thermometer does not *cause* temperature.

Figure 2: Schematic diagrams of the gas-mode Michelson Interferometer, with beam splitter/mirror at A and mirrors at B and C mounted on arms from A, with the arms of equal length L_0 when at rest. D is the detector screen. In Top the interferometer is at rest in space. In Bottom the instrument and gas are moving through 3-space with speed v_R parallel to the AB arm. Interference fringes are observed at D when mirrors B and C are not exactly perpendicular - the Hick's effect. As the interferometer is rotated in the plane shifts of these fringes are seen in the case of absolute motion, but only if the apparatus operates in a gas. By measuring fringe shifts the speed v_R may be determined.

$$\Delta t = k^2 \frac{L v_P^2}{c^3} \cos(2(\theta - \psi)), \qquad (13)$$

where ψ specifies the direction of $\mathbf{v}(\mathbf{r},t)$ projected onto the plane of the interferometer, giving projected speed v_P, relative to the local meridian, and where

$k^2 = (n^2 - 2)(n^2 - 1)$, with n the refractive index. Neglect of the absolute motion relativistic Fitzgerald-Lorentz contraction effect gives $k^2 \approx n^3 \approx 1$ for gases, which is essentially the Newtonian theory that Michelson used.

We derive the calibration constant k^2 for the Michelson interferometers in the case of Lorentzian Relativity. The two arms are constructed to have the same lengths when they are physically parallel to each other. For convenience assume that the value L_0 of this length refers to the lengths when at rest wrt space The Fitzgerald-Lorentz effect is that the arm AB parallel to the direction of motion is shortened to

$$L_\parallel = L_0 \sqrt{1 - \frac{v_R^2}{c^2}} \qquad (14)$$

where v_R is the lengthwise speed of the arm relative to space. In SR v_R is the speed relative to the observer, who is presumably at rest wrt the arms, then $v_R = 0$ and there is no arm contraction effect.

For later reference we also give the time dilation expression for physical clocks:

$$\tau = T \sqrt{1 - \frac{v_R^2}{c^2}} \qquad (15)$$

where τ is the elapsed time given by the clock, for an actual time interval T.

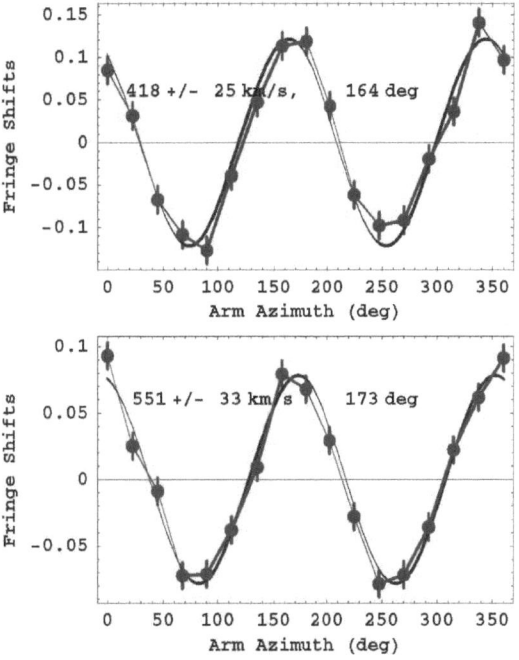

Figure 3: Top: Typical Miller data from 1925/26 gas-mode Michelson interferometer, from averaging 20 360° rotations, (Miller, 1933). Bottom: Data from Michelson-Morley 1887 gas-mode interferometer, from averaging 6 360° rotations. In both plots the non-orthogonal term and temperature drift effects have been removed from the data, after a least squares best fit using the full detector theory derived in the text. This reduced data then shows an impressive agreement with the $\cos(2(\theta - \psi))$ form.

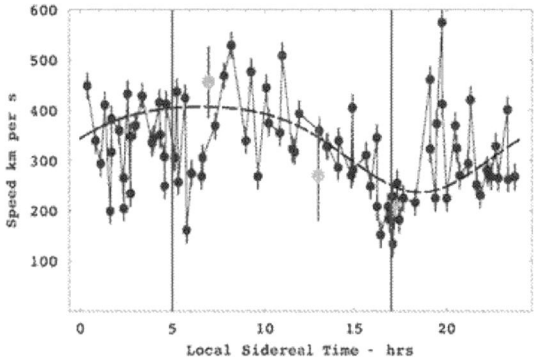

Figure 4: Speeds v_P, of the 3-space velocity $\mathbf{v}$ projected onto the horizontal plane of the Miller gas-mode Michelson interferometer located atop Mt.Wilson, plotted against local sidereal time in hours, for a composite day, with data collected over a number of days in September 1925. The data shows considerable fluctuations, from hour to hour, and also day to day, as this is a composite day. The dashed curve shows the non-fluctuating best-fit variation over one day, as the earth rotates, causing the projection onto the plane of the interferometer of the velocity of the average direction of the space

flow to change. The maximum projected speed of the curve is 417 km/s (using the STP air refractive index of $n=1.00029$, and the min/max occur at approximately 5hrs and 17hrs local sidereal time (Right Ascension). Note that the Cassini flyby in August, [16], gives a RA=5.15^h, close to the RA apparent in the above plot. The data points, with error bars, at 7^h and 13^h are from the Michelson-Morley 1887 data, from averaging (excluding only the July 8 data for 7^h because it has poor S/N). The fiducial time lines are at 5^h and 17^h. The speed fluctuations are seen to be much larger than the statistically determined errors, confirming the presence of turbulence in the 3-space flow, i.e gravitational waves, as first seen in the Michelson-Morley experiment.

For simplicity here we take the motion of the detector to be parallel to the arm AB. Following Fig.2 let the time taken for light to travel from $A \rightarrow B$ be t_{AB} and that from $B \rightarrow A$ be t_{BA}, where V is the speed of light relative to the gas, which is moving with the detector. We shall also neglect the Fresnel drag effect, so $V=c/n$. Then $Vt_{AB}=L_\| + v_R t_{AB}$ and $Vt_{BA}=L_\| - v_R t_{BA}$.

$$t_{ABA} = t_{AB} + t_{BA} = \frac{L_\|}{V-v_R} + \frac{L_\|}{V+v_R}$$

$$= \frac{2L_0 V \sqrt{1 - \frac{v_R^2}{c^2}}}{V^2 - v_R^2}. \qquad (16)$$

For the other arm, with no contraction in its length,

$$\left(Vt_{AC} \right)^2 = L_0^2 + \left(v_R t_{AC} \right)^2$$

$$t_{AC} = \frac{L_0}{\sqrt{V^2 - v_R^2}}, \quad t_{ACA} = 2t_{AC} = \frac{2L_0}{\sqrt{V^2 - v_R^2}}, \qquad (17)$$

giving finally for the travel time difference for the two arms

$$\Delta t = \frac{2L_0 V \sqrt{1 - \frac{v_R^2}{c^2}}}{V^2 - v_R^2} - \frac{2L_0}{\sqrt{V^2 - v_R^2}}. \qquad (18)$$

Now trivially $\Delta t=0$ if $v_R=0$, but also $\Delta t=0$ when $v_R \neq 0$ but only if $V=c$, *viz* vacuum. This then would result in a null result on rotating the apparatus. Hence the null result of the Michelson apparatus is only for the special case of light travelling in vacuum. However if the apparatus is immersed in a gas then $V<c$ and a non-null effect is expected on rotating the apparatus, since now

$\Delta t \neq 0$. It is essential then in analysing data to correct for this refractive index effect. Putting $V=c/n$ in (18) we find, for $v_R \square V$ and when $n \approx 1$, that

$$\Delta t = n(n^2 - 1) \frac{L_0 v_R^2}{c^3}. \qquad (19)$$

However if the data is analyzed not using the Fitzgerald-Lorentz contraction (14), then, as done in the old analyses, the estimated time difference is

$$\Delta t = \frac{2L_0 V}{V^2 - v_R^2} - \frac{2L_0}{\sqrt{V^2 - v_R^2}}, \qquad (20)$$

which again for $v_R \square V$ gives

$$\Delta t = n^3 \frac{L_0 v_R^2}{c^3}. \qquad (21)$$

With Fresnel drag and $n \approx 1$, the sign of Δt is reversed. Symmetry arguments easily show that when rotated we obtain a $\cos(2\theta)$ factor.

However the above analysis does not correspond to how the interferometer is actually operated. That analysis does not actually predict fringe shifts, for the field of view would be uniformly illuminated, and the observed effect would be a changing level of luminosity rather than fringe shifts. As Michelson and Miller knew, the mirrors must be made slightly non-orthogonal with the degree of non-orthogonality determining how many fringe shifts were visible in the field of view. Experimenting with this effect determines a comfortable number of fringes: not too few and not too many. The non-orthogonality reduces the symmetry of the device, and instead of having period of 180° the symmetry now has a period of 360°, so that we must add the extra term $a\cos(\theta - \beta)$ in

$$\Delta t = k^2 \frac{L(1+e\theta)v_P^2}{c^3} \cos\left(2(\theta - \psi) \right) + $$
$$a(1+e\theta)\cos(\theta - \beta) + f, \qquad (22)$$

The factor $1+e\theta$ models the temperature effects, namely that as the arms are uniformly rotated, one rotation taking several minutes, there will also be a temperature induced change in the length of the arms. If the temperature effects are linear in time, as they would be for short time intervals, then they are linear in θ. In the non-orthogonality term the parameter a is proportional to the length of the arms, and so also has the temperature factor. The term f simply models any offset effect. Michelson and Morley and Miller took these two effects into account when analyzing his data.

The interferometers are operated with the arms horizontal. Then θ is the azimuth of one arm relative to the local meridian, while ψ is the azimuth of the absolute motion velocity projected onto the plane of the interferometer, with projected component v_p. Here the Fitzgerald-Lorentz contraction is a real dynamical effect of absolute motion, unlike the Einstein space-time view that it is merely a space-time perspective artifact, and whose magnitude depends on the choice of observer. The instrument is operated by rotating at a rate of one rotation over several minutes, and observing the shift in the fringe pattern through a telescope during the rotation. Then fringe shifts from six (Michelson and Morley) or twenty (Miller) successive rotations are averaged to improve the signal to noise ratio, and the average sidereal time noted. Some examples are shown in Fig.2, and illustrate the incredibly clear signal. The ongoing claim that the Michelson-Morley experiment was a null experiment is disproved. Fig.4 shows data from these two experiments over a 24hr sidereal day. The large fluctuations are gravitational wave effects, and have been seen in all experiments that detected light speed anisotropy.

8. DeWitte RF Coaxial Cable Detector

The enormously significant 1991 DeWitte double one-way 1st order in v/c experiment successfully measured the anisotropy of the speed of RF EM waves using clocks at each end of the RF coaxial cables [12,21]. The technique uses rotation of the coaxial cables, by means of the earth rotation, to permit extraction of the EM speed anisotropy, despite the clocks not being synchronized. Data from this 1st order in v/c experiment agrees with the speed and direction of the anisotropy results from all the other experiments reported herein.

Figure 5: Schematic layout for measuring the one-way speed of light in either free-space, optical fibers or RF coaxial cables, without requiring the synchronization of the clocks C_1 and C_2. Here τ is the, initially unknown, offset time between the clocks. Times t_A and t_B are true times, without clock offset and clock transport effects, while $T_A = t_A$, $T_B = t_A + \tau$ and $T'_B = t'_B + \tau + \Delta\tau$ are clock readings. $V(v\cos(\theta))$ is the speed of EM radiation wrt the apparatus before rotation, and $V(v\cos(\theta - \Delta\theta))$ after rotation, **v** is the velocity of the apparatus through space in direction θ relative to the apparatus before rotation, **u** is the velocity of transport for clock C_2, and $\Delta\tau < 0$ is the net slowing of clock C_2 from clock transport, when apparatus is rotated through angle $\Delta\theta > 0$. Note that **v·u** > 0.

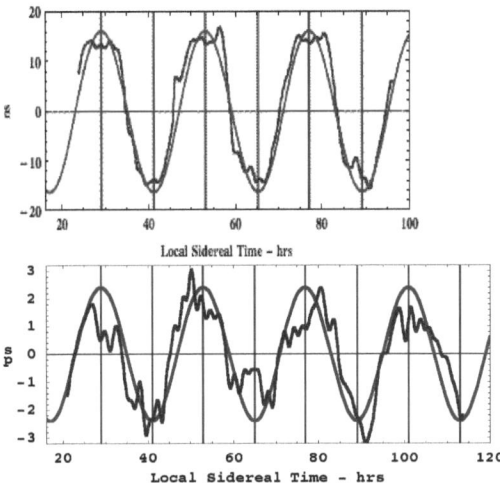

Figure 6: Top: Data from the 1991 DeWitte NS RF coaxial cable experiment, L=1.5km, using the arrangement shown in Fig.5, with a 2nd RF coaxial cable carrying a signal in the reverse direction. The vertical red lines are at RA=5^h. DeWitte gathered data for 178 days, and showed that the crossing time tracked sidereal time, and not local solar time, see Fig.7. DeWitte reported that $v \approx 500 km/s$. If the full Fresnel drag effect is included no effect would have been seen. Bottom: Dual RF coaxial cable detector data from May 2009, Cahill [20], using the technique in Fig.13 with L=20m. NASA Spacecraft Doppler shift data predicts Dec=-77°, v=480km/s, giving a sidereal dynamic range of 5.06ps, very close to that observed. The vertical red lines are at RA=5^h. In both data sets we see the earth sidereal rotation effect together with significant wave/turbulence effects

Fig.5 shows the arrangement for measuring the one-way speed of light, either in vacuum, a dielectric, or RF coaxial cable. It is usually argued that one-way speed of light measurements are not possible because the clocks C_1 and C_2 cannot be synchronized. However this is false. An important effect that needs to be included is the clock offset effect caused by transport, when the

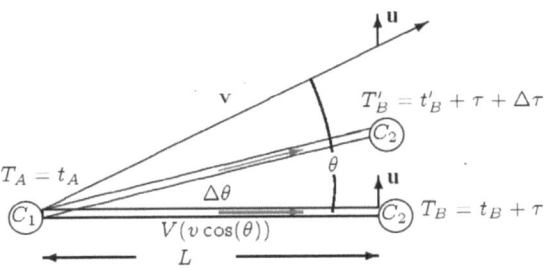

apparatus is rotated in this case, but most significantly the Fresnel drag effect is not present in RF coaxial cables, at low RF frequencies. In Fig.5 the actual travel time $t_{AB} = t_B - t_A$ from A to B, as distinct from the clock indicated travel time $T_{AB} = T_B - T_A$, is determined by

$$V(v\cos(\theta))t_{AB} = L + v\cos(\theta)t_{AB} \tag{23}$$

where the 2nd term comes from the end B moving an additional distance $v\cos(\theta)t_{AB}$ during time interval t_{AB}. With Fresnel drag $V(v) = \frac{c}{n} + v\left(1 - \frac{1}{n^2}\right)$, when V and v are parallel, and where n is the dielectric refractive index. Then

$$t_{AB} = \frac{L}{V(v\cos(\theta)) - v\cos(\theta)} = \frac{nL}{c} + \frac{v\cos(\theta)L}{c^2} + .. \tag{24}$$

However if there is no Fresnel drag effect, $V = c/n$, as is the case in RF coaxial cables, then we obtain

$$t_{AB} = \frac{L}{V(v\cos(\theta)) - v\cos(\theta)} = \frac{nL}{c} + \frac{v\cos(\theta)Ln^2}{c^2} + .. \tag{25}$$

It would appear that the two terms in (24) or (25) can be separated by rotating the apparatus, giving the magnitude and direction of **v**. However it is $T_{AB} = T_B - T_A$ that is measured, and not t_{AB}, because of an unknown fixed clock offset τ, as the clocks are not *a priori* synchronized, and as well an angle dependent clock transport offset $\Delta\tau$, at least until we can establish clock synchronizations, as explained below. Then the clock readings are $T_A = t_A$ and $T_B = t_B + \tau$, and $T_B' = t_B' + \tau + \Delta\tau$, where $\Delta\tau$ is a clock offset that arises from slowing of clock C_2 as it is transported during the rotation through angle $\Delta\theta$, see Fig.5.

The clock transport offset $\Delta\tau$ follows from the clock motion effect [21]

$$\Delta\tau = dt\sqrt{1 - \frac{(\mathbf{v} + \mathbf{u})^2}{c^2}} - dt\sqrt{1 - \frac{\mathbf{v}^2}{c^2}} = -dt\frac{\mathbf{v} \cdot \mathbf{u}}{c^2} + .. \tag{26}$$

when clock C_2 is transported at velocity **u** over time interval dt, compared to C_1. Now $\mathbf{v} \cdot \mathbf{u} = vu\sin(\theta)$ and $dt = L\Delta\theta/u$. Then the change in T_{AB} from this small rotation is, using (25) for the case of no Fresnel drag,

$$\Delta T_{AB} = -\frac{v\sin(\theta)Ln^2\Delta\theta}{c^2} + \frac{v\sin(\theta)L\Delta\theta}{c^2} + ... \tag{27}$$

as the clock transport effect appears to make the clock-determined travel time longer (2nd term). Integrating we get

$$T_B - T_A = \frac{nL}{c} + \frac{v\cos(\theta)L(n^2 - 1)}{c^2} + \tau, \tag{28}$$

where τ is now the constant offset time. The $v\cos(\theta)$ term may be separated by means of the angle dependence. Then the value of τ may be determined, and the clocks synchronized.

However if the propagation medium is liquid, or dielectrics such as glass and optical fibers, the Fresnel drag effect is present, and we then use (24), and not (25). Then in (28) we need make the replacement $n \rightarrow 1$, and then the 1st order in v/c term vanishes. However, in principle, separated clocks may be synchronized by using RF coaxial cables.

The DeWitte $L = 1.5$km 5MHz RF coaxial cable experiment, in Brussels in 1991, was a double 1st order in v/c detector, using the scheme in Fig.5, but employing a 2nd RF coaxial cable for the opposite direction, giving clock difference $T_C - T_D$, to cancel temperature effects, and also used 3 Caesium atomic clocks at each end. The orientation was NS and rotation was achieved by that of the earth. Then

$$T_{AB} - T_{CD} = \frac{2v\cos(\theta)L(n^2 - 1)}{c^2} + 2\tau \tag{29}$$

For a horizontal detector the dynamic range of $\cos(\theta)$ is $2\sin(\lambda)\cos(\delta)$, caused by the earth rotation, where λ is the latitude of the detector location and δ is the declination of **v**. The value of τ may be determined and the clocks synchronised. Some of DeWiite's data and results are in Figs.6 and 7. We see that DeWitte's RF EM speed anisotropy experiment is consistent with other experiments, and also shows significant fluctuations.

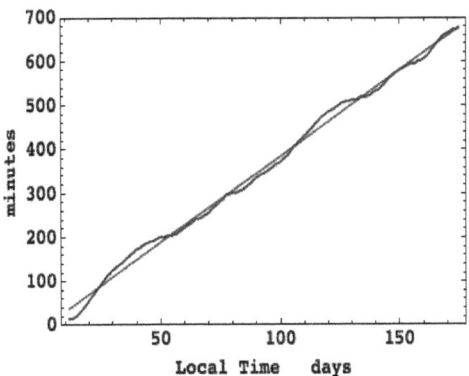

Figure 7: DeWitte collected data over 178 days and demonstrated that the zero crossing time, see Fig.6, tracked sidereal time and not local solar time. The plot shows the negative of the drift in the crossing time vs local solar time, and has a slope, determined by the best-fit straight line, of -3.918 minutes per day, compared to the actual average value of -3.932 minutes per day, the difference between a sidereal day and a solar day.

9. Earth Flyby RF Doppler Shifts: 3-Space Flow

The motion of spacecraft relative to the earth are measured by observing the direction and Doppler shift of the transponded RF EM transmissions. This gives another technique to determine the speed and direction of the dynamical 3-space as manifested by the light speed anisotropy [16]. The repeated detection of the anisotropy of the speed of light has been, until recently, ignored in analysing the Doppler shift data, causing the long-standing anomalies in the analysis [1]. The use of the Minkowski-Einstein choice of time and space coordinates does not permit the analysis of these Doppler anomalies, as they mandate that the speed of the EM waves be invariant.

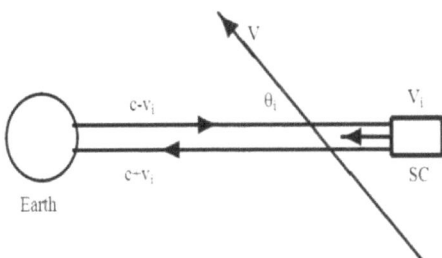

Figure 8: Asymptotic flyby configuration in earth frame-of-reference, with spacecraft (SC) approaching Earth with velocity $\mathbf{V}_i$.

The departing asymptotic velocity will have a different direction but the same speed, as no force other than conventional Newtonian gravity is assumed to be acting upon the SC. The dynamical 3-space velocity is $\mathbf{v}(\mathbf{r},t)$, though taken to be time independent during the Doppler shift measurement, which causes the outward EM beam to

have speed $c-v_i(r)$, and inward speed $c+v_i(r)$, where $v_i(r)=v(r)\cos(\theta_i)$, with θ_i the angle between $\mathbf{v}$ and $\mathbf{V}$. A similar description applies to the departing SC, labeled $i{\to}f$.

Because we shall be extracting the earth inflow effect we need to take account of a spatially varying, but not time-varying, 3-space velocity. In the earth frame of reference, see Fig.8, and using clock times from earth-based clocks, let the transmitted signal from earth have frequency f. The time for one RF maximum to travel distance D to SC from earth is, see Fig.9,

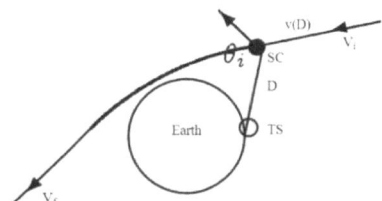

Figure 9: Spacecraft (SC) earth flyby trajectory, with initial and final asymptotic velocity $\mathbf{V}$, differing only by direction. The Doppler shift is determined from Fig.8 and (43). The 3-space flow velocity at the location of the SC is $\mathbf{v}$. The line joining Tracking Station (TS) to SC is the path of the RF signals, with length D. As SC approaches earth $\mathbf{v}(D)$ changes direction and magnitude, and hence magnitude of projection $v_i(D)$ also changes, due to earth component of 3-space flow and also because of RF direction to/from Tracking Station. The SC trajectory averaged magnitude of this earth in-flow is determined from the flyby data and compared with theoretical prediction.

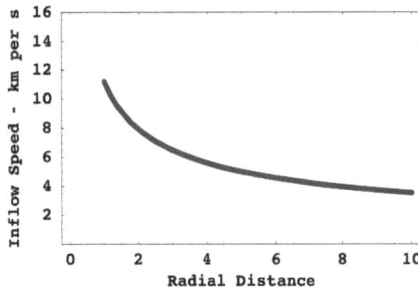

Figure 10: Predicted Earth 3-space inflow speed vs distance from earth in earth radii $v=\sqrt{2GM/R}$, plotted only for $R>1.0$. Combining the NASA/JPL optical fiber RA determination and the flyby Doppler shift data has permitted the determination of the angle- and distance-averaged inflow speed, to be 12.4 ± 5km/s

$$t_1 = \int_0^D \frac{dr}{c - v_i(r)} \tag{30}$$

The next RF maximum leaves time $T=1/f$ later and arrives at SC at time, taking account of SC motion,

$$t_2 = T + \int_0^{D-VT} \frac{dr}{c - v_i(r)} \tag{31}$$

The period at the SC of the arriving RF is then

$$T' = t_2 - t_1 = T + \int_D^{D-VT} \frac{dr}{c - v_i(r)} \approx \frac{c - v_i(D) - V}{c - v_i(D)} T \tag{32}$$

Essentially this RF is reflected[§] by the SC. Then the 1st RF maximum takes time to reach the earth

$$t_1' = -\int_D^0 \frac{dr}{c + v_i(r)} \tag{33}$$

and the 2nd RF maximum arrives at the later time

$$t_2' = T' - \int_{D-VT'}^0 \frac{dr}{c + v_i(r)}. \tag{34}$$

Then the period of the returning RF at the earth is

$$T'' = t_2' - t_1', \tag{35}$$

$$= T' + \int_D^{D-VT'} \frac{dr}{c + v_i(r)}, \tag{36}$$

$$\approx \frac{c + v_i(D) - V}{c + v_i(D)} T', \tag{37}$$

Then overall we obtain the return frequency to be

$$f'' = \frac{1}{T''} = \frac{c + v_i(D)}{c + v_i(D) - V} \frac{c - v_i(D)}{c - v_i(D) - V} f \tag{38}$$

Ignoring the projected 3-space velocity $v_i(D)$, that is, assuming that the speed of light is invariant as per the usual literal interpretation of the Einstein 1905 light speed postulate, we obtain instead

$$f'' = \frac{c^2}{(c-V)^2} f. \tag{39}$$

The use of (39) instead of (38) is the origin of the putative anomalies. Expanding (39) we obtain

$$\frac{\Delta f}{f} = \frac{f'' - f}{f} = \frac{2V}{c} \tag{40}$$

However expanding (38) we obtain, for the same Doppler shift,

$$\frac{\Delta f}{f} = \frac{f'' - f}{f} = \left(1 + \frac{v(D)^2}{c^2}\right) \frac{2V}{c} + \dots \tag{41}$$

It is the prefactor to $2V/c$ missing from (40) that explains the spacecraft Doppler anomalies, and also permits yet another determination of the 3-space

velocity $\mathbf{v}(D)$ at the location of the SC. The published data does not give the Doppler shifts as a function of SC location, so the best we can do at present is to use a SC trajectory-averaged $v(D)$, namely $\overline{v}_i$ and $\overline{v}_f$ for the incoming and outgoing trajectories, as further discussed below.

From the observed Doppler shift data acquired during a flyby, and then best fitting the trajectory, the asymptotic hyperbolic speeds $V_{i\infty}$ and $V_{f\infty}$ are inferred from (40), but incorrectly so, as in Anderson [1]. These inferred asymptotic speeds may be related to an inferred asymptotic Doppler shift

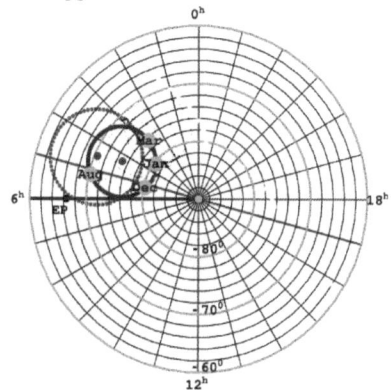

Figure 11: South celestial pole region. The dot at RA=4.3^h, Dec=75° S, and with speed 486km/s, is the direction of motion of the solar system through space determined from NASA spacecraft earth-flyby Doppler shifts, revealing the EM radiation speed anisotropy. The thick circle centered on this direction is the observed velocity direction for different days of the year, caused by earth orbital motion and sun 3-space inflow. The corresponding results from the Miller gas-mode interferometer are shown by 2nd dot and its aberration circle. For December 8, 1992, the velocity is RA=5.2^h, Dec=80°S, speed 491km/s.

$$\frac{\Delta f_{i\infty}}{f} = \frac{f_\infty' - f}{f} = \frac{2V_{i\infty}}{c} + \dots \tag{42}$$

which from (41) gives

$$V_{i\infty} \equiv \frac{\Delta f_{i\infty}}{f} \cdot \frac{c}{2} = \left(1 + \frac{\overline{v}_i^2}{c^2}\right) V + \dots \tag{43}$$

where V is the actual asymptotic speed. Similarly after the flyby we obtain

$$V_{f\infty} \equiv \frac{\Delta f_{f\infty}}{f} \cdot \frac{c}{2} = \left(1 + \frac{\overline{v}_f^2}{c^2}\right) V + \dots \tag{44}$$

[§]In practice a more complex protocol is used.

and we see that the "asymptotic" speeds $V_{i\infty}$ and $V_{f\infty}$ must differ, as indeed reported in Anderson, 2008 [1]. We then obtain the expression for the so-called flyby anomaly

$$\Delta V_\infty = V_{f\infty} - V_{i\infty} = \frac{\overline{v}_f^2 - \overline{v}_i^2}{c^2} \qquad (45)$$

where here $V \approx V_\infty$ to sufficient accuracy, where V_∞ is the average of $V_{i\infty}$ and $V_{f\infty}$. The existing data on **v** permits *ab initio* predictions for ΔV_∞. As well a separate least-squares-fit to the individual flybys permits the determination of the average speed and direction of the 3-space velocity, relative to the earth, during each flyby. These results are all remarkably consistent with the data from the various laboratory experiments that studied **v**. We now indicate how $\overline{v}_i$ and $\overline{v}_f$ were parametrized during the best-fit to the flyby data. $\mathbf{v}_{galactic} + \mathbf{v}_{sun} - \mathbf{v}_{orbital}$ is taken as constant during each individual flyby, with $\mathbf{v}_{sun}$ inward towards the sun, with value 42 km/s, and $\mathbf{v}_{orbital}$ as tangential to earth orbit with value 30 km/s - consequentially the directions of these two vectors changed with day of each flyby. This linear superposition is only approximate, Cahill [14]. The earth inflow $\mathbf{v}_{earth}$ was taken as radial and of an unknown fixed trajectory-averaged value. So the averaged direction but not the averaged speed varied from flyby to flyby, with the incoming and final direction being approximated by the (α_i, δ_i) and (α_f, δ_f) asymptotic directions. The predicted theoretical variation of $v_{earth}(R)$ is shown in Fig.10.

This results in the plot in Fig.11 and the earth in-flow speed determination. The results are in remarkable agreement with the results from Miller, showing the extraordinary skill displayed by Miller in carrying out his massive interferometer experiment and data analysis in 1925/26. The only effect missing from the Miller analysis is the spatial in-flow effect into the sun, which affected his data analysis. Miller obtained a galactic flow direction of $\alpha = 4.52$ hrs, $\delta = -70.5^o$, compared to that obtained herein from the NASA data of $\alpha = 4.29$ hrs, $\delta = -75.0^o$, which differ by only $\approx 5^o$.

As well the flyby Doppler shifts show considerable fluctuations. We conjecture that these are gravitational wave effects, although no analysis has been done to characterise these fluctuations.

The numerous EM anisotropy experiments discussed herein demonstrate that a dynamical 3-space exists, and that the speed of the earth wrt this space exceeds 1 part in 1000 of c, namely a large effect. Not surprisingly this has indeed been detected many times over the last 127 years. The speed of ~ 500 km/s means that earth based clocks experience a real, so-called, time dilation effect from (15) of approximately 0.12s per day compared to cosmic time. However clocks may be corrected for this clock dilation effect because their speed v though space, which causes their slowing, is measurable by various experimental methods. This means that the absolute or cosmic time of the universe is measurable. This very much changes our understanding of time. However because of the inhomogeneity of the earth 3-space in-flow component the clock slowing effect causes a differential effect for clocks at different heights above the earth's surface. It was this effect that Pound and Rebka reported in 1960 using the Harvard tower [35]. Consider two clocks at heights h_1 and h_2, with $h = h_2 - h_1$, then the frequency differential follows from (15)

$$\frac{\Delta f}{f} = \sqrt{1 - \frac{v^2(h_2)}{c^2}} - \sqrt{1 - \frac{v^2(h_1)}{c^2}}, \qquad (46)$$

$$\approx \frac{v^2(h_1) - v^2(h_2)}{2c^2} + ..., \qquad (47)$$

$$= \frac{1}{2c^2}\frac{dv^2(r)}{dr}h + ..., \qquad (48)$$

$$= \frac{g(r)h}{c^2} + ..., \qquad (49)$$

$$= -\frac{\Delta\Phi}{c^2} + ..., \qquad (50)$$

where Φ is the so-called 'gravitational potential', and with $\mathbf{v}.\nabla\mathbf{v} = \nabla\left(\frac{v^2}{2}\right)$ for zero vorticity $\nabla\times\mathbf{v} = \mathbf{0}$, and ignoring any time dependence of the flow, and where finally, $\Delta\Phi$ is the change in the gravitational potential. The actual process here is that, say, photons are emitted at the top of the tower with frequency f and reach the bottom detector with the same frequency f - there is no change in the frequency. This follows from (32) but with now $V=0$ giving $T=T'$. However the bottom clock is running slower because the speed of space there is faster, and so this clock determines that the falling photon has a higher frequency, i.e. appears blue shifted. The opposite effect is seen for upward travelling photons, namely an apparent red shift as observed by the top clock. In practice the Pound-Rebka experiment used motion induced Doppler shifts to make these measurements using the Mössbauer effect. The overall conclusion is that Pound and Rebka measured the derivative of v^2 wrt height, whereas herein we have measured that actual speed, but averaged wrt the SC trajectory measurement protocol. It is important to note that the so-called "time dilation" effect is really a "clock slowing" effect - clocks are simply slowed by their movement through 3-space. The Gravity Probe A experiment also studied the clock slowing effect, though again interpreted differently therein, and again complicated by additional Doppler effects.

The Cosmic Microwave Background (CMB) velocity is often confused with the Absolute Motion (AM) velocity or light-speed anisotropy velocity as determined in the experiments discussed herein. However these are unrelated and in fact point in very different directions, being almost at 90^0 to each other, with the CMB velocity being 369 km/s in direction $(\alpha=11.2^h, \delta=-7.22^0)$.

The CMB velocity is obtained by defining a frame of reference in which the thermalised CMB 3^0K radiation is isotropic, that is by removing the dipole component, and the CMB velocity is the velocity of the Earth in that frame. The CMB velocity is a measure of the motion of the solar system relative to the last scattering surface (a spherical shell) of the universe some 13.4Gyrs in the past. The concept here is that at the time of decoupling of this radiation from matter that matter was on the whole, apart from small observable fluctuations, on average at rest with respect to the 3-space. So the CMB velocity is not motion with respect to the *local* 3-space now; that is the AM velocity. Contributions to the AM velocity would arise from the orbital motion of the solar system within the Milky Way galaxy, which has a speed of some 250 km/s, and contributions from the motion of the Milky Way within the local cluster, and so on to perhaps super clusters, as well as flows of space associated with gravity in the

Milky Way and local galactic cluster etc. The difference between the CMB velocity and the AM velocity is explained by the spatial flows that are responsible for gravity at the galactic scales.

10. Dual RF Coaxial Cable Detector

The Dual RF Coaxial Cable Detector, Cahill [20] exploits the Fresnel drag anomaly, in that there is no Fresnel drag effect in RF coaxial cables, at low enough frequencies, see Fig.12.

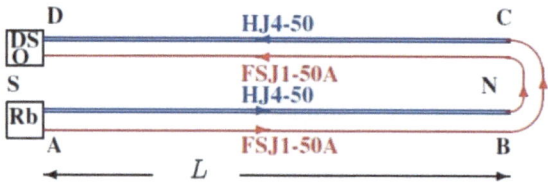

Figure 12: Because Fresnel drag is absent in RF coaxial cables this dual cable setup, using one clock, is capable of detecting the absolute motion of the detector wrt to space, revealing the sidereal rotation effect as well as wave/turbulence effects. In the 1st trial of this detector this arrangement was used, with the cables laid out on a laboratory floor, and results are shown in Figs 6, bottom. In the new design the cables in each circuit are configured into 8 loops, as in Fig.14, giving $L=8\times1.85\text{m}=14.8\text{m}$. In comparison with data from spacecraft earth-flyby Doppler shifts, Cahill [], this experiments confirms that there is no Fresnel drag effect in RF coaxial cables. In Cahill [] a version with optical fibers in place of the HJ4-50 coaxial cables was used, see Fig.18. There the optical fiber has a Fresnel drag effect while the coaxial cable did not. In that experiment optical-electrical converters were used to modulate/demodulate infrared light.

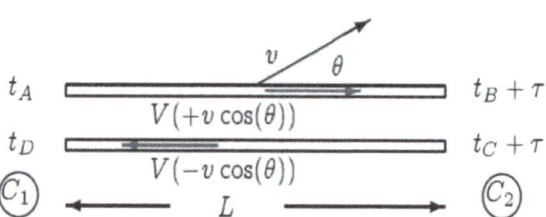

Figure 13: Schematic layout for measuring the one-way speed of EM waves in RF coaxial cables, V is the speed of EM radiation wrt the apparatus, with or without the Fresnel drag and v is the speed of the apparatus through space, in direction θ. Times here refer to absolute times.

Figure 14: Photograph of the RF coaxial cables arrangement, based upon 16 ×1.85m lengths of phase stabilized Andrew HJ4-50 coaxial cable. These are joined to 16 lengths of phase stabilized Andrew FSJ1-50A cable, in the manner shown schematically in Fig.12. The 16 HJ4-50 coaxial cables have been tightly bound into a 4×4 array, so that the cables, locally, have the same temperature, with cables in one of the circuits embedded between cables in the 2nd circuit. This arrangement of the cables permits the cancellation of temperature differential effects in the cables. A similar array of the smaller diameter FSJ1-50A cables is located inside the conduit boxes.

Fig.13 shows the arrangement for measuring the one-way speed EM waves in RF coaxial cable. The actual travel time t_{AB} from A to B is determined by

$$V(v\cos(\theta))t_{AB} = L + v\cos(\theta)t_{AB} \qquad (51)$$

where the 2nd term comes from the end B moving an additional distance $v\cos(\theta)t_{AB}$ during time interval t_{AB}. Then

$$t_{AB} = \frac{L}{V(v\cos(\theta)) - v\cos(\theta)} = \frac{nL}{c} + \frac{v\cos(\theta)L}{c^2} + .. \qquad (52)$$

$$t_{CD} = \frac{L}{V(v\cos(\theta)) + v\cos(\theta)} = \frac{nL}{c} - \frac{v\cos(\theta)L}{c^2} + .. \qquad (53)$$

on using the Fresnel effect, and expanding to 1st order in v/c. However if there is no Fresnel drag effect then we obtain

$$t_{AB} = \frac{L}{V(v\cos(\theta)) - v\cos(\theta)} = \frac{nL}{c} + \frac{v\cos(\theta)Ln^2}{c^2} + .. \qquad (54)$$

$$t_{CD} = \frac{L}{V(v\cos(\theta)) + v\cos(\theta)} = \frac{nL}{c} - \frac{v\cos(\theta)Ln^2}{c^2} + .. \qquad (55)$$

The important observation is that the v/c terms are independent of the dielectric refractive index n in (52) and (53), but have an n^2 dependence in (54) and (55), in the absence of the Fresnel drag effect. Then from (54) and (55) the round trip travel time is, see Fig.12,

$$t_{AB} + t_{CD} = \frac{(n_1+n_2)L}{c} + \frac{v\cos(\theta)L(n_1^2-n_2^2)}{c^2} + .. \qquad (56)$$

where n_1 and n_2 are the effective refractive indices for the two different RF coaxial cables, with two separate circuits to reduce temperature effects. Shown in Fig.14 is a photograph. The Andrews Phase Stabilised FSJ1-50A has n_1=1.19, while the HJ4-50 has n_2=1.11.

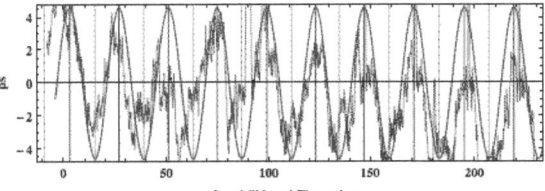
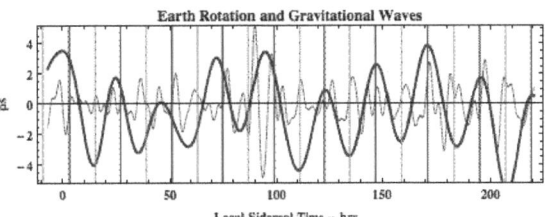

Figure 15: Top: Travel time differences (ps) between the two coaxial cable circuits in Fig.12, orientated NS and horizontal, over 9 days (March 4-12, 2012, Adelaide) plotted against local sidereal time. Sinewave, with dynamic range 8.03ps, is prediction for sidereal effect from flyby Doppler shift data for RA=2.75^h (shown by vertical fiducial lines), Dec=-76.6^o, and with speed 499.2km/s, see Table 1. Data shows sidereal effect and significant wave/turbulence effects. Bottom: Data filtered into two frequency bands 3.4×10^{-3}mHz$<f<0.018$mHz ($81.4h>T>15.3h$) and 0.018mHz$<f<0.067$mHz ($15.3h>T>4.14h$), showing more clearly the earth rotation sidereal effect (plus very low frequency waves) and the turbulence without the sidereal effect. Frequency spectrum of top data is shown in Fig.16.

One measures the travel time difference of two RF 10MHz signals from a Rubidium frequency standard (Rb) with a Digital Storage Oscilloscope (DSO). In each circuit the RF signal travels one-way in one type of coaxial cable, and returns via a different kind of coaxial cable. Two circuits are used so that temperature effects cancel - if a temperature change alters the speed in one type of cable, and so the travel time, that travel time change is the same in both circuits, and cancels in the difference. The travel time difference of the two circuits at the DSO is

$$\Delta t = \frac{2v\cos(\theta)L(n_1^2-n_2^2)}{c^2} + .. \qquad (57)$$

If the Fresnel drag effect occurred in RF coaxial cables, we would use (54) and (55) instead, and then the $n_1^2-n_2^2$ term is replaced by 0, i.e. there is no 1st order term in v. The preliminary layout for this detector used cables laid out as in Fig.12, and the data is shown in Fig.6. In the compact design the Andrew HJ4-50 cables are cut into 8 × 1.85m shorter lengths in each circuit, corresponding to a net length of $L=8\times1.85=14.8$m, and the Andrew FSJ1-50A cables are also cut, but into longer lengths to enable joining. However the curved parts of the Andrew FSJ1-50A cables contribute only at 2nd order in v/c. The apparatus was horizontal and orientated NS, and used the rotation of the earth to change the angle θ. The dynamic range of $\cos(\theta)$, caused by the earth rotation only, is again $2\sin(\lambda)\cos(\delta)$, where $\lambda=-35^{o}$ is the latitude of Adelaide. Inclining the detector at angle λ removes the earth rotation effect, as now the detector arm is parallel to the earth's spin axis, permitting a more accurate characterisation of the wave effects.

The cable travel times and the DSO phase measurements still have a temperature dependence, and these effects are removed from the data, rather than attempt to maintain a constant temperature, which is impractical because of the heat output of the Rb clock and DSO. The detector was located in a closed room in which the temperature changed slowly over many days, with variations originating from changing external weather driven temperature changes. The temperature of the detector was measured, and it was assumed that the timing errors were proportional to changes in that one measured temperature. These timing errors were some 30ps, compared to the true signal of some 8ps. Because the temperature timing errors are much larger, the temperature induced $\Delta t=a+b\Delta T$ was fitted to the timing data, and the coefficients a and b determined. Then this Δt time series was subtracted from the data, leaving the actual required phase data. This is particularly effective as the temperature variations had a distinctive time signature.

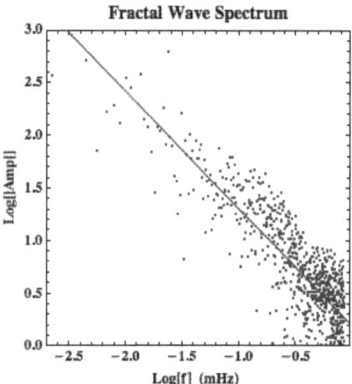

Figure 16: Log-Log plot of the data (top) in Fig.15, with the straight line being $A\propto1/f$, indicating a $1/f$ fractal wave spectrum. The interpretation for this is the 3-space structure shown in Fig.17.

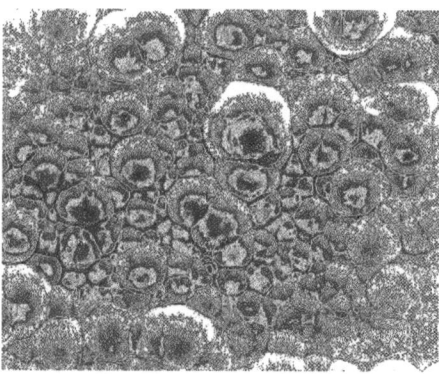

Figure 17: Representation of the fractal wave data as a revealing the fractal textured structure of the 3-space, with cells of space having slightly different velocities, and continually changing, and moving wrt the earth with a speed of ~500km/s.

The phase data, after removing the temperature effects, is shown in Fig.15 (top), with the data compared with predictions for the sidereal effect only from the flyby Doppler shift data. As well that data is separated into two frequency bands (bottom), so that the sidereal effect is partially separated from the gravitational wave effect, *viz* 3-space wave turbulence. Being 1st order in v/c it is easily determined that the space flow is from the southerly direction. (Miller, 1933) reported the same sense, i.e. the flow is essentially from S to N, though using a 2nd order detector that is more difficult to determine. The frequency spectrum of this data is shown in Fig.16, revealing a fractal $1/f$ form. This implies the fractal structure of the 3-space indicated in Fig.17

11. Optical Fiber RF Coaxial Cable Detector

An earlier 1st order in v/c gravitational wave detector design is shown in Fig.18, Cahill [8,9], with some data shown in Fig.19. Only now is it known why that detector also worked, namely that there is a Fresnel drag effect in the optical fibers, but not in the RF coaxial cable. Then the travel time difference, measured at the DSO, is given by

$$\Delta t = \frac{2v\cos(\theta)L(n_1^2-1)}{c^2}+..\quad(58)$$

where n_1 is the effective refractive index of the RF coaxial cable. Again the data is in remarkable agreement with the flyby and other detections of **v**.

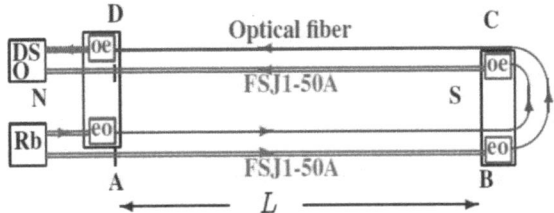

Figure 18: Layout of the optical fiber - coaxial cable detector, with L=5.0m. 10MHz RF signals come from the Rubidium atomic clock (Rb). The Electrical to Optical converters (EO) use the RF signals to modulate 1.3μm infrared signals that propagate through the single-mode optical fibers. The Optical to Electrical converters (OE) demodulate that signal and give the two RF signals that finally reach the Digital Storage Oscilloscope (DSO), which measures their phase difference. The key effects are that the propagation speeds through the coaxial cables and optical fibers respond differently to their absolute motion through space, with no Fresnel drag in the coaxial cables, and Fresnel drag effect in the optical fibers. Without this key difference this detector does not work.

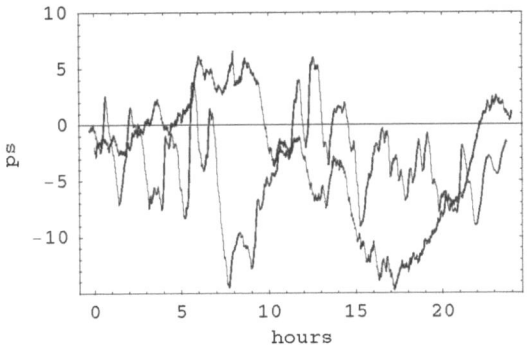

Figure 19: Phase difference (ps), with arbitrary zero, versus local time data plots from the Optical Fiber - Coaxial Cable Detector, see Fig.18 and Cahill [8], showing the sidereal time effect and significant wave/turbulence effects.. The plot with the most easily identified minimum at ~17hrs local Adelaide time is from June 9, 2006, while the other plot with the minimum at ~8.5hrs local time is from August 23, 2006. We see that the minimum has moved forward in time by approximately 8.5 hrs. The expected sidereal shift for this 65 day difference, without wave effects, is 4.3 hrs, to which must be added another ~1h from the aberration effects shown in Fig11, giving 5.3hrs, in agreement with the data, considering that on individual days the min/max fluctuates by ±2hrs. This sidereal time shift is a critical test for the detector. From the flyby Doppler data we have for August RA=5^h, Dec=-70°, and speed 478km/s, giving a predicted sidereal effect dynamic range to be 8.6ps, very close to that observed

The Dual RF Coaxial Cable Detector exploits the Fresnel drag anomaly in RF coaxial cables, *viz* the drag effect is absent in such cables, for reasons unknown, and this 1st order in v/c detector is compact, robust and uses one clock. This anomaly now explains the operation of the Optical-Fiber - Coaxial Cable Detector,

and permits a new calibration. These detectors have confirmed the absolute motion of the solar system and the gravitational wave effects seen in the earlier experiments of Michelson-Morley, Miller, DeWitte. Most significantly these experiments agree with one another, and with the absolute motion velocity vector determined from spacecraft earth-flyby Doppler shifts. The observed significant wave/ turbulence effects reveal that the so-called "gravitational waves" are easily detectable in small-scale laboratory detectors, and are considerably larger than those predicted by GR. These effects are not detectable in vacuum-mode Michelson terrestrial interferometers, nor by their analogue vacuum-mode resonant cavity experiments.

The Dual RF Coaxial Cable Detector permits a detailed study and characterization of the wave effects, and with the detector having the inclination equal to the local latitude the earth rotation effect may be removed, as the detector is then parallel to the earth's spin axis, enabling a more accurate characterization of the wave effects. The major discovery arising from these various results is that 3-space is directly detectable and has a fractal textured structure. This and numerous other effects are consistent with the dynamical theory for this 3-space. We are seeing the emergence of fundamentally new physics, with space being a non-geometrical dynamical system, and fractal down to the smallest scales describable by a classical velocity field, and below that by quantum foam dynamics Cahill [6].

12. Quantum Zener Diode Detectors

When extending the Dual RF Coaxial Cable Detector experiment to include one located in London, in addition to that located in Adelaide, an analysis of the measured DSO internal noise in each identically setup instrument was undertaken, when the extensive RF coaxial cable array was replaced by short leads. This was intended to determine the S/N ratio for the joint Adelaide-London experiment. Surprisingly the internal noise was found to be correlated, with the noise in the London DSO being some 13 to 20 seconds behind the Adelaide DSO** noise, see Fig.20. The correlation data had a phase that tracked sidereal time, meaning that the average direction was approximately fixed wrt the galaxy, but with extensive fluctuations as well from the gravitational wave/turbulence effect, that had been seen

---

** LeCroy WaveRunner 6051A DSOs were used.

in all previous experiments. The explanation for this DSO effect was not possible as the DSO is a complex instruments, and which component was responding to the passing space fluctuations could not be determined. But the correlation analysis did demonstrate that not all of the internal noise in the DSO was being caused solely by some random process intrinsic to the instrument.

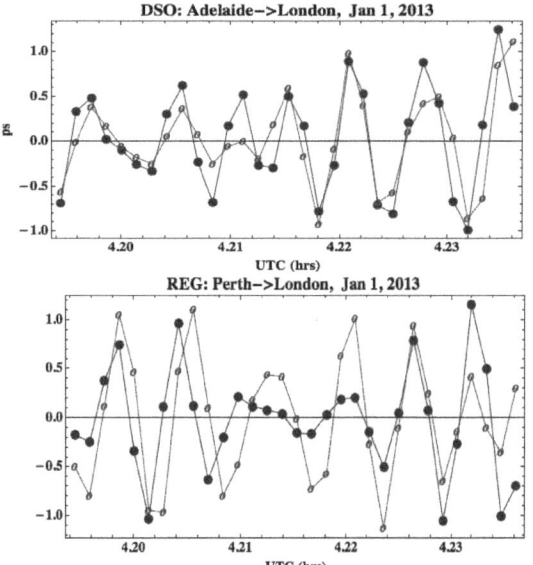

Figure 20: Correlations in band-passed Adelaide-London DSO data (top) and Perth (Australia)-London REG data (bottom), for January 1, 2013, with London data (open dots) advanced by 15s in both cases, over the same 200s time interval. The data points are at 5s intervals. The REG data was recorded every 1s, and has been averaged to 5s intervals for ease of comparison with DSO data. The UTC time at all detectors was determined using internet timing applications, which have ms precision.

The travel time delay $\tau(t)$ was determined by computing the correlation function

$$C(\tau,t)= \int_{t-T}^{t+T} dt' S_1(t'-\tau/2)S_2[t'+\tau/2]e^{-a(t'-t)^2} \tag{59}$$

for the two detector signals $S_1(t)$ and $S_2(t)$. Here $2T=200s$ is the time interval used, about UTC time t. The gaussian term ensures the absence of end-effects. Maximising $C(\square,t)$ wrt τ gives $\tau(t)$ - the delay time vs UTC t, and plotted in Figs. 21 and 22, where the data has been binned into 1hr time intervals, and the rms also shown. The speed and direction, over a 24hr period, was determined by fitting the time delay data using

$$\tau= \frac{\mathbf{R}\cdot\mathbf{v}}{\mathbf{v}^2}, \tag{60}$$

where $\mathbf{R}$ is the Adelaide-London spatial separation vector, and $\mathbf{v}(\theta,\delta)$ is the 3-space velocity vector, parametrized by a speed, RA and Declination. This expression assumes a plane wave form for the gravitational waves. The $\tau(t)$ delay times show large fluctuations, corresponding to fluctuations in speed and/or direction, as also seen in the data in Fig.4, and also a quasi-periodicity, as seen in Fig.20. Then only minimal travel times, $10s<\tau<22s$, were retained. Correlations, as shown in Fig.20, are not always evident, and then the correlation function $C(\tau,t)$ has a low value. Only $\tau(t)$ data from high values of the correlation function were used. The absence of correlations at all times is expected as the London detector is not directly "downstream" of the Adelaide detector, and so a fractal structure to space, possessing a spatial inhomogeneity, bars ongoing correlations, and as well the wave structure will evolve during the travel time. Fig.20 shows examples of significant correlations in phase and amplitude between all four detectors, but with some mismatches. The approximate travel time of 15s in Fig.20 at ~4.2hrs UTC is also apparent in Fig.21, with the top figure showing the discovery of the correlations from the two DSO separated by a distance $R\approx12160km$. That the internal "noise" in these DSO is correlated is a major discovery.

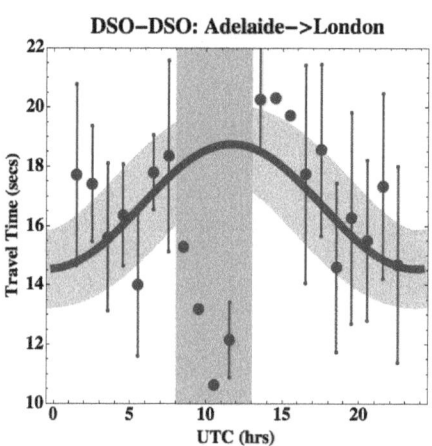

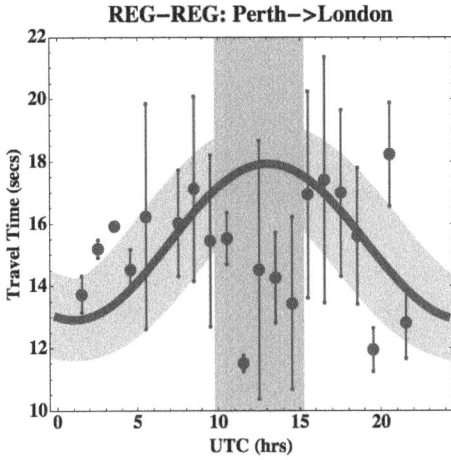

Figure 21: Travel times from DSO-DSO Adelaide-London data (top), and REG-REG Perth-London data (bottom) from correlation analysis using (59). The data in each 1 hr interval has been binned, and the average and rms shown. The thick (red line) shows best fit to data using plane wave travel time predictor, (60), but after excluding those data points between 8 and 13hrs UTC (top) and 10 and 15hrs UTC (bottom), indicated by vertical band. Those data points are not consistent with the plane wave modelling, and suggest a scattering process when the waves pass deeper into the earth, see fig.23. The Perth-London phase is retarded wrt Adelaide-London phase by ~1.5hrs, consistent with Perth being 1.5hrs west of Adelaide. The Adelaide-London data gives speed = 512 km/s, RA = 4.8 hrs, Dec = 83°S, and the Perth-London data gives speed = 528 km/s, RA = 5.3 hrs, Dec = 81°S. The broad band tracking the best fit line is for +/- 1 sec fluctuations, corresponding to speed fluctuation of +/- 17km/s. Actual fluctuations are larger than this, as 1st observed by Michelson-Morley and by Miller, see Fig.4.

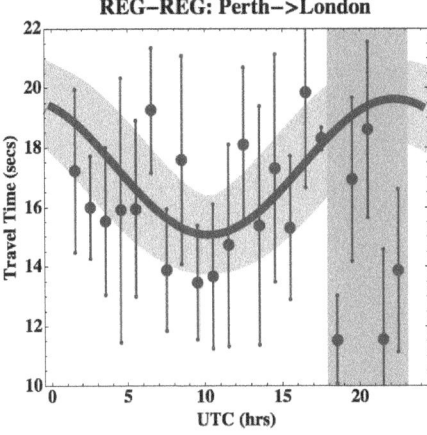

Figure 22: Travel times from REG-REG Perth-London data for August 1, 2012. The data in each 1 hr interval has been binned, and the average and rms shown. The thick line shows best fit to data using plane wave travel time predictor, (60), but after excluding those data points between 18 and 23hrs UTC, indicated by vertical band. Those data points are not consistent with the plane wave modelling. This data gives speed = 471 km/s, RA = 4.4 hrs, Dec = 82°S. The change in phase of the maximum of the data, from UTC= 22+/-2 hr, for

August 1, 2012, to UTC = 12+/- 2 hr for January 2013 (Fig.21), but with essentially the same RA, illustrates the sidereal effect: the average direction of the space flow is fixed wrt to the stars, apart from the earth-orbit aberration effect, Fig.11.

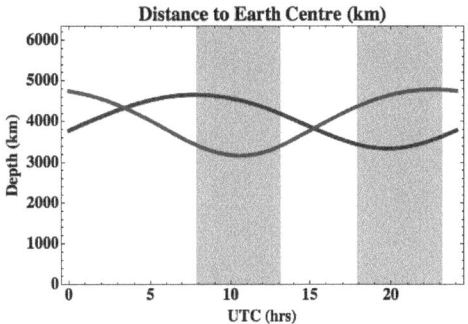

Figure 23: Given measured space velocity, plots show maximum earth penetration depth of space detected by London detectors for Adelaide→London, Jan1, 2013 and Perth→London, August 1, 2012 , revealing that the anomalous scattering occurs when deeper depths are "traversed". The vertical shadings correspond to those in Fig.21 (top) and Fig.22.

There are much simpler devices that were discovered to also display time delayed correlations over large distances: these are the Random Number Generators (RNG) or Random Event Generators (REG). There are various designs available from manufacturers, and all claim that these devices manifest hardware random quantum processes, as they involve the quantum to classical transition when a measurements, say, of the quantum tunnelling of electrons through a nanotechnology potential barrier, ~10nm thickness, is measured by a classical/macroscopic system. According to the standard interpretation of the quantum theory, the collapse of the electron wave function to one side or the other of the barrier, after the tunnelling produces a component on each side, is purely a random event, internal to the quantum system. However this interpretation had never been tested experimentally. Guided by the results from the DSO correlated-noise effect, the data from two REGs, located in Perth and London, was examined. The data[††] showed the same correlation effect as observed in the DSO experiments, see Figs. 20-22. However REGs typically employ a XOR gate that

---

[††]The data is from the GCP international network: http://teilhard.global-mind.org/.

The data[‡‡] showed the same correlation effect as observed in the DSO experiments, see Figs. 20-22. However REGs typically employ a XOR gate that produces integer valued outputs with a predetermined statistical form.

To study the zener diode tunnelling currents without XOR gate intervention two collocated zener diode circuits were used to detect highly correlated tunnelling currents, Figs.24 and 25. When the detectors are separated by ~0.25m in NS direction, phase differences ~0.5µs were observed and dependent on relative orientation. So this zener diode circuit forms a very simple and cheap nanotechnology quantum detector for gravitational waves.

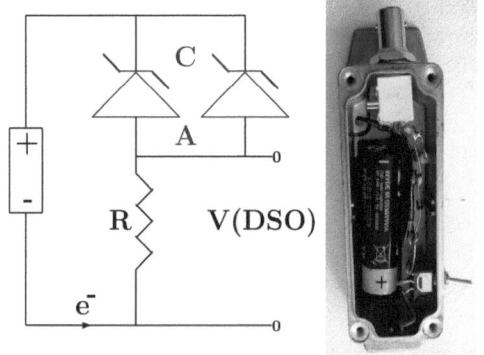

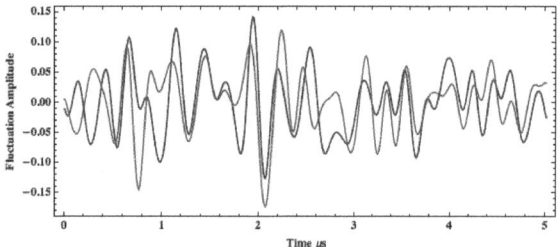

Figure 24: Left: Circuit of Zener Diode Gravitational Wave Detector, showing 1.5V AA battery, two 1N4728A zener diodes operating in reverse bias mode, and having a Zener voltage of 3.3V, and resistor R= 10KΩ. Voltage V across resistor is measured and used to determine the space driven fluctuating tunnelling current through the zener diodes. Correlated currents from two collocated detectors are shown in Fig.25. Right: Photo of detector with 6 zener diodes in parallel.

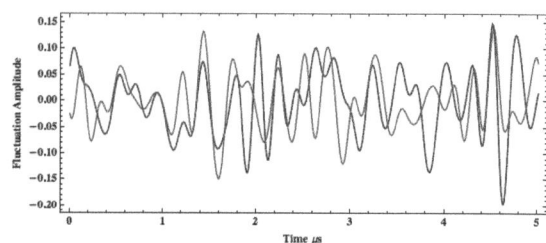

Figure 25: Top: Current fluctuations from two collocated zener diode detectors, Fig.24 (separated by 3-4 cm in EW direction due to box size), revealing strong correlations. The small separation may explain slight differences, revealing a structure to space at very small distances. Bottom: Correlations when detectors separated NS by approximately 25 cm, and with N detector signal advanced by 0.5 µs, and then showing strong correlations. This time delay effect reveals space traveling from S to N at a speed of approximately 500km/s. Fig.26 shows plot of correlation function $C(\tau,t)$, with time delay □ expressed as a speed over a distance of 25 cm.

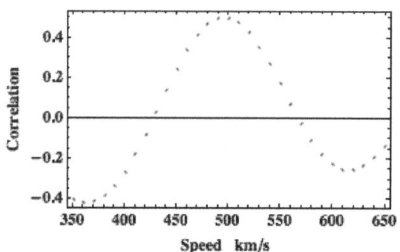

Figure 26: Correlation function $C(\tau,t)$, (59), with time delay □ expressed as a speed over a distance of 25 cm, for the data shown in Fig.25, Bottom. t is the time of observation, which is not relevant in this test case. This plot reveals a speed of 500±25 km/s.

13. Dynamical 3-Space

If Michelson and Morley had more carefully presented their pioneering data physics would have developed in a very different direction. Even by 1925/26 Miller, a junior colleague of Michelson, was repeating the gas-mode interferometer experiment, and by not using Newtonian mechanics to attempt a calibration of the device, rather by using the earth aberration effect which utilised the earth orbital speed of 30km/s to set the calibration constant, although that also entailed false assumptions. The experimental data reveals the existence of a dynamical space. It is a simple matter to derive the dynamics of space, and the emergence of gravity as a quantum matter effect.

Physics must employ a covariance formulation, in the sense that ultimately predictions are independent of observers, and that there must also be a relativity principle that relates observational data by different observers. We assume then that space has a structure whose movement, wrt an observer, is described by a velocity field, $\mathbf{v}(\mathbf{r},t)$, at the classical physics level, at a location $\mathbf{r}$ and time t, as defined by the observer. In particular the space coordinates $\mathbf{r}$ define an embedding space, which herein we take to be Euclidean. At a deeper level space is probably a fractal quantum foam, which is only approximately embeddable in a 3-dimensional space at a coarse-grained level, Cahill [6,15,17]. This embedding space has no ontological existence - it is not real. Ironically Newton took this space to be real but unobservable, and so a different concept, and so excluding the possibility that gravity was caused by an accelerating space. It is assumed that different observers, in relative uniform motion, relate their description of the velocity field by means of the Galilean Relativity Transformation for positions and velocities. It is usually argued that the Galilean Relativity Transformations were made redundant and in error by the Special Relativity Transformations. However this is not so - there exist an exact linear mapping between Galilean Relativity and Special Relativity (SR), differing only by definitions of space and time coordinates Cahill [12,13]. This implies that the so-called Special Relativity (SR) relativistic effects are not actual dynamical effects - they are purely areifacts of a peculiar choice of space and time coordinates. In particular Lorentz symmetry is merely a consequence of this choice of space and time coordinates, and is equivalent to Galilean symmetry. Nevertheless Lorentz symmetry remains valid, even though a local preferred frame of reference exists. Lorentz Relativity, however, goes beyond Galilean Relativity in that the limiting speed of systems wrt to the local space causes various so-called relativist effects, such as length contractions and clock dilations.

The Euler covariant constituent acceleration $\mathbf{a}(\mathbf{r},t)$ of space is then defined by

$$\mathbf{a} = \lim_{\Delta t \to 0} \frac{\mathbf{v}(\mathbf{r}+\mathbf{v}(\mathbf{r},t)\Delta t, t+\Delta t) - \mathbf{v}(\mathbf{r},t)}{\Delta t}$$

$$= \frac{\partial \mathbf{v}}{\partial t} + (\mathbf{v}\cdot\nabla)\mathbf{v}$$

which describes the acceleration of a constituent element of space by tracking its change in velocity. This means that space has a (quantum) structure that permits its velocity to be defined and detected, which experimentally has been done. We assume here that the flow has zero vorticity $\nabla\times\mathbf{v}=\mathbf{0}$, and then the flow is

determined by a scalar function $\mathbf{v}=\nabla u$. We then need one scalar equation to determine the space dynamics, which we construct by forming the divergence of $\mathbf{a}$. The inhomogeneous term then determines a dissipative flow caused by matter, expressed as a matter density, and where the coefficient turns out to be Newton's gravitational constant,

$$\nabla\cdot\left(\frac{\partial \mathbf{v}}{\partial t}+(\mathbf{v}\cdot\nabla)\mathbf{v}\right)=-4\pi G\rho(\mathbf{r},t) \qquad (61)$$

Note that even a time independent matter density or even the absence of matter can be associated with a time-dependent flow. In particular this dynamical space in the absence of matter has an expanding universe solution. Substituting the Hubble form $\mathbf{v}(\mathbf{r},t)=H(t)\mathbf{r}$, and then using $H(t)=\dot{a}(t)/a(t)$, where $a(t)$ is the scale factor of the universe, we obtain the solution $a(t)=t/t_0$, where t_0 is the age of the universe, since by convention $a(t_0)=1$.

Then computing the magnitude-redshift function $\mu(z)$, we obtain excellent agreement with the supernova red-shift data, Cahill and Rothall [26].

This equation follows essentially from covariance and dimensional analysis. For a spherically symmetric matter distribution, of total mass M, and a time-independent spherically symmetric flow we obtain from the above, and external to the sphere of matter, the acceleration of space

$$\mathbf{v}(r)=\sqrt{\frac{2GM}{r}}\,\hat{r} \quad \text{giving } \mathbf{a}(r)=\frac{2GM}{r^2}\,r \qquad (62)$$

which is the inverse square law. Newton applied such an acceleration to matter, not space, and which Newton invented directly by examining Kepler's planetary motion laws, but which makes no mention of what is causing the acceleration of matter, although in a letter in 1675 to Oldenburg, Secretary of the Royal Society, and later to Robert Boyle, he speculated that an undetectable ether flow through space may be responsible for gravity. Here, however, the inverse square law emerges from the Euler constituent acceleration, which imposes a space self-interaction. At the surface of the earth the in-flow speed is 11km/s, and the sun in-flow speed at 1AU is 42km/s, with both detected Cahill, [15].

While the above 3-space dynamical equation followed from covariance and dimensional analysis, this derivation is not complete yet. One can add additional terms with the same order in speed and spatial derivatives, and which cannot be *a priori* neglected. These developments have been extensively tested with experiments and observations, Cahill [6,10,15,24], Rothall and Cahill [36,37] and Cahill and Kerrigan [25].

14. Quantum Matter and 3-Space: Emergent Gravity

We now derive, uniquely, how quantum matter responds to the dynamical 3-space. This gives the 1st derivation of the phenomenon of gravity, and reveals this to be a quantum matter wave refraction effect. For a free-fall quantum system with mass m the Schrödinger equation is uniquely generalised, Cahill [11], with the new terms required to maintain that the motion is intrinsically wrt the 3-space, and not wrt the embedding space, and that the time evolution is unitary

$$i\hbar \frac{\partial}{\partial t}\Psi(r,t) = \left[\frac{-\hbar^2}{2m}\nabla^2 - i\hbar(\mathbf{v}\cdot\nabla + \frac{1}{2}\nabla\cdot\mathbf{v})\right]\Psi(r,t)$$

The space and time coordinates $\{t,x,y,z\}$ ensure that the separation of a deeper and unified process into different classes of phenomena - here a dynamical 3-space (quantum foam) and a quantum matter system, is properly tracked and connected. As well the same coordinates may be used by an observer to also track the different phenomena. A quantum wave packet propagation analysis gives the matter acceleration $\mathbf{g}=d^2<\mathbf{r}>/dt^2$ induced by wave refraction to be

$$\mathbf{g}=\frac{\partial\mathbf{v}}{\partial t}+(\mathbf{v}.\nabla)\mathbf{v}+(\nabla\times\mathbf{v})\times\mathbf{v}_R+...$$

$$\mathbf{v}_R(\mathbf{r}_0(t),t)=\mathbf{v}_0(t)-\mathbf{v}(\mathbf{r}_0(t),t),$$

where $\mathbf{v}_R$ is the velocity of the wave packet relative to the 3-space, and where $\mathbf{v}_O$ and $\mathbf{r}_O$ are the velocity and position relative to the observer. The last term generates the Lense-Thirring effect as a vorticity driven effect. In the limit of zero vorticity we obtain that the quantum matter acceleration is the same as the 3-space acceleration: $\mathbf{g}=\mathbf{a}$. This confirms that the new physics is in agreement with Galileo's observations that all matter falls with the same acceleration. Using arcane language this amounts to a derivation of the Weak Equivalence Principle.

Significantly the quantum matter 3-space-induced 'gravitational' acceleration also follows from maximising the elapsed proper time wrt the quantum matter wave-packet trajectory $\mathbf{r}_o(t)$, Cahill [7],

$$\tau = \int dt\sqrt{1-\frac{\mathbf{v}_R^2(\mathbf{r}_0(t),t)}{c^2}} \qquad (63)$$

which entails that matter has a maximum speed of c wrt to space, and not wrt an observer. This maximisation ensures that quantum waves propagating along neighbouring paths are in phase - the condition for a classical trajectory. This gives

$$\mathbf{g}=\frac{\partial\mathbf{v}}{\partial t}+(\mathbf{v}\cdot\nabla)\mathbf{v}+(\nabla\times\mathbf{v})\times\mathbf{v}_R-\frac{\mathbf{v}_R}{1-\frac{\mathbf{v}_R^2}{c^2}}\frac{1}{2}\frac{d}{dt}\left(\frac{\mathbf{v}_R^2}{c^2}\right)+... \quad (64)$$

and then taking the limit $v_R/c\to 0$ we recover the non-relativistic limit, above. This shows that (i) the matter 'gravitational' geodesic is a quantum wave refraction effect, with the trajectory determined by a Fermat maximum proper-time principle, and (ii) that quantum systems undergo a local time dilation effect. The last, relativistic, term generates the planetary precession effect. If clocks are forced to travel different trajectories then the above predicts different evolved times when they again meet - this is the Twin Effect, which now has a simple and explicit physical explanation - it is an absolute motion effect, meaning motion wrt space itself. This elapsed proper time expression invokes Lorentzian relativity, that the maximum speed is c wrt to space, and not wrt the observer, as in Einstein SR. The differential proper time (63) has the form

$$c^2 d\tau^2 = c^2 dt^2 - (d\mathbf{r}-\mathbf{v}(\mathbf{r},t)dt)^2 = g_{\mu\nu}dx^\mu dx^\nu$$

which defines an induced metric for a curved spacetime manifold. However this has no ontological significance, and the metric is not determined by GR.

15. Electromagnetic Radiation and Dynamical Space

We must generalise the Maxwell equations so that the electric and magnetic fields are excitations within the dynamical 3-space, and not of the embedding space. The minimal form in the absence of charges and currents is

$$\nabla\times\mathbf{E}=-\mu_0\left(\frac{\partial\mathbf{H}}{\partial t}+\mathbf{v}.\nabla\mathbf{H}\right), \quad \nabla.\mathbf{E}=\mathbf{0},$$

$$\nabla\times\mathbf{H}=\varepsilon_0\left(\frac{\partial\mathbf{E}}{\partial t}+\mathbf{v}.\nabla\mathbf{E}\right), \quad \nabla.\mathbf{H}=\mathbf{0}$$

which was first suggested by Hertz [29], but with $\mathbf{v}$ then being only a constant vector field, and not interpreted as a moving space effect. As easily determined the speed of EM radiation is now $c=1/\sqrt{\mu_0\varepsilon_0}$ with respect to the 3-space, and not wrt an observer in motion through the 3-space. The Michelson-Morley 1887 experiment 1st detected this anisotropy effect, as have numerous subsequent experiments. A time-dependent and/or inhomogeneous velocity field causes the refraction of EM radiation. This can be computed by using the Fermat least-time approximation - the opposite of that for quantum matter. This ensures that EM waves along neighboring paths are in phase. Then an EM ray path $\mathbf{r}(t)$ is determined by minimizing the elapsed travel time:

$$T = \int_{s_i}^{s_f} \frac{ds\,|\,d\mathbf{r}/ds\,|}{|\,c\hat{\mathbf{v}}_R(s) + \mathbf{v}(\mathbf{r}(s)t(s))\,|},$$

with $\quad \mathbf{v}_R = \dfrac{d\mathbf{r}}{dt} - \mathbf{v}(\mathbf{r}(t),t) \qquad$ by varying both $\mathbf{r}(s)$ and $t(s)$, finally giving $\mathbf{r}(t)$. Here s is an arbitrary path parameter, and $c\hat{\mathbf{v}}_R$ is the velocity of the EM radiation wrt the local 3-space, namely c. The denominator is the speed of the EM radiation wrt the observer's Euclidean spatial coordinates. This equation may also be used to calculate the gravitational lensing by black holes, filaments, Cahill [18], and by ordinary matter, using the appropriate 3-space velocity field. It produces the measured light bending by the sun.

16 Conclusions

Herein is reviewed extensive experimental evidence that a dynamical 3-space is the foundation process of reality and, in particular, that the speed of light is not invariant, as claimed in the space-time theory that has dominated academic physics since 1905. This dynamical space passes the earth with a speed ~500km/s from a near southerly direction, and the first detection of that is now understood to go back to the Michelson-Morley experiment of 1887, now that it is understood that the original report in 1887 was flawed by a lack of understanding of how the interferometer should have been calibrated. That misunderstanding led to the development of the space-time model. More recently a variety of experimental techniques have been developed, with the latest using the Zener diode quantum detectors, that permit the detection of this space, which the data shows has a fractal structure. The theory for this dynamical space has been found by generalizing Newtonian gravity by first converting that to a velocity field formalism, which then immediately permitted a generalization that did not alter the inverse square law outside of spherically symmetric masses. This dynamical theory has permitted the resolution of numerous anomalies in physics, g decreasing more

slowly down boreholes than predicted by NG or GR, inconsistent laboratory measurements of G, flat rotation plots for spiral galaxies, star dynamics near the center of the Milky Way central black hole, effects of a earth centered black hole, which causes the g anomaly, and its effect on the generation of matter within the earth via intense 3-space fluctuations [17]. [19], cosmic filaments [18] and networks of black holes connected by cosmic filaments [25], uniformly expanding universe [26] without the need for 'dark matter' nor 'dark energy', which were merely fix-ups for General Relativity, which failed to explain any of the above anomalies. Determination of location in space and in time requires the use of no-Lorentz Relativity, and involves correction for the effects of absolute motion through space upon clocks and rods. Generalizing the quantum theory to include this dynamical space led to gravity being an emergent quantum phenomenon, being caused by the refraction of quantum matter waves by the dynamical space: the unification of gravity and quantum physics. The dynamical space theory has no measure of energy or energy content, contrary to the usual Zero Point Energy notion, but its interaction with quantum matter generates energy; this is the violation of the conservation of energy principle, and had previously been discovered by N. Tesla, T.T. Brown, W. Reich, H. Moray, and others. This effect is the basis for the Zener diode quantum detection of the dynamical space, and also correlations between dynamical space fluctuations and Solar flare rates, which leads to a new explanation of the correlation between Solar flare counts and the Earth's climate [27], and also the anisotropic Brownian motion of colloidal particle droplets in water [28]. We note that dynamical 3-space is not an aether substance in a inactive geometrical space, but a complex quantum dynamical system, which at a sufficiently large scale permits a geometrical and velocity field description [6], and with quantum behavior arising from a pattern recognizing activity within a neural network like system, with such patterns being semantic information within that system.

A special thank you to Martin Kokus for organizing the NPA meeting in Baltimore in 2014.

References

1.	Anderson J.D., Campbell J.K., Ekelund J.E., Ellis J. and Jordan J.F. Anomalous Orbital-Energy Changes Observed during Spacecraft Flybys

2.   Brown H.R. The Origins of Length Contraction; I The Fitzgerald-Lorentz Deformation Hypothesis, Am. J. Phys., v. 69, 1044-1054 (2001).

3.   Brown H.R. Physical Relativity: Space-Time Structure from a Dynamical Perspective, Clarendon Press Oxford (2005).

4.   Brown H.R. and Pooley O. Minkowski Space-Time: a Glorious Non-Entity, in The Ontology of Spacetime, Dieks, ed. Elsevier, 67-89 (2006).

5.   Cahill R.T. and Kitto K. Michelson-Morley Experiments Revisited, Apeiron, 10(2),104-117 (2003).

6.   Cahill R.T. Process Physics: From Information Theory to Quantum Space and Matter, Nova. Sci. Pub. NY (2005).

7.   Cahill R.T. The Michelson and Morley 1887 Experiment and the Discovery of Absolute Motion, *Progress in Physics*, 3, 25-29, (2005).

8.   Cahill R.T. A New Light-Speed Anisotropy Experiment: Absolute Motion and Gravitational Waves Detected, *Progress in Physics*, 4, 73-92 (2006).

9.   Cahill R.T. Absolute Motion and Gravitational Wave Experiment Results, Contribution to Australian Institute of Physics National Congress, Brisbane, Paper No. 202 (2006).

10. Cahill R.T. 3-Space Inflow Theory of Gravity: Boreholes, Black holes and the Fine Structure Constant, *Progress in Physics*, 2, 9-16 (2006).

11. Cahill R.T. Dynamical Fractal 3-Space and the Generalized Schrödinger Equation: Equivalence Principle and Vorticity Effects, Progress in Physics, 1, 27-34 (2006).

12. Cahill R.T. The Roland De Witte 1991 Experiment, *Progress in Physics*, 3, 60-65 (2006).

13. Cahill R.T., *Progress in Physics*, 4, 19-24 (2008).

14. Cahill R.T. The Dynamical Velocity Superposition Effect in the Quantum-Foam Theory of Gravity, in Relativity, Gravitation, Cosmology: New Developments, Dvoeglazov V., ed., Nova Science Pub., New York (2009).

15. Cahill R.T. Dynamical 3-Space: A Review, in Ether Space-time and Cosmology: New Insights into a Key Physical Medium, Duffy and Levy, eds., Apeiron, 135-200 (2009).

16. Cahill R.T. Combining NASA/JPL One-Way Optical-Fiber Light Speed Data with Spacecraft Earth-Flyby Doppler-Shift Data to Characterise 3-Space Flow, *Progress in Physics*, 4, 50-64 (2009).

17. Cahill R.T. Dynamical 3-Space: Emergent Gravity, in Should the Laws of Gravity be Reconsidered? Munera, H.A., ed., Apeiron, Montreal, 363-376 (2011).

18. Cahill R.T. *Progress in Physics*, 2, 44-51 (2011).

19. Cahill R.T. Dynamical 3-Space and the Earth's Black Hole :An Expanding Earth Mechanism. In G Scalera, E Boschi and S Cwojdziński, eds. The Earth Expansion Evidence - A Challenge for Geology, Geophysics and Astronomy. 1 ed. Rome, Italy: ARACNE editrice. 37th Workshop of the International School of Geophysics. Erice, Italy 2011, 185-196 (2012).

20. Cahill R.T. Characterisation of Low Frequency Gravitational Waves from Dual RF Coaxial-Cable Detector: Fractal Textured Dynamical 3-Space, *Progress in Physics*, 3, 3-10, (2012).

21. Cahill R.T. One-Way Speed of Light Measurements Without Clock Synchronisation, Progress in Physics, 3, 43-45 (2012).

22. Cahill R.T. Dynamical 3-Space: Neo-Lorentz Relativity, Physics International 4(1), 60-72. doi:10.3844/pisp.2013.60.72 Published Online 4 (1) (http://www.thescipub.com/pi.toc) (2013).

23. Cahill R.T. Nanotechnology Quantum Detectors for Gravitational Waves: Adelaide to London Correlations Observed, Progress in Physics, 4, 57-62 (2013).

24. Cahill R.T. Discovery of Dynamical 3-Space: Theory, Experiments and Observations - A Review, American Journal of Space Science, doi:10.3844 /ajssp.2013.77.93 Published Online 1(2) (http://www.thescipub.com/ajss.toc) (2013).

25. Cahill R.T. and Kerrigan D., *Progress in Physics* 4, 79-82 (2011).

26. Cahill R.T. and Rothall D. Discovery of Uniformly Expanding Universe, Progress in Physics, 1, 63-68 (2012).

27. Cahill R.T. Solar Flare Five-Day Predictions from Quantum Detectors of Dynamical Space Fractal Flow Turbulence: Gravitational Wave Diminutionand Earth Climate Cooling, Progress in Physics, v.10, 236-242 (2014).

28. Cahill R.T. Dynamical 3-Space: Energy Non-Conservation, Anisotropic Brownian Motion

Experiment and Ocean Temperatures, in preparation (2015).

29. Hertz H. On the Fundamental Equations of Electro-Magnetics for Bodies in Motion, Wiedemann's Ann. 41, 369; (1890). Electric Waves, Collection of Scientific Papers, Dover Pub., New York (1962).

30. Lorentz H.A. Electromagnetic Phenomena in System Moving with any Velocity Smaller than that of Light , Proc. of the Royal Netherlands Academy of Arts and Sciences, v. 6, 809-831 (1904).

31. Lorentz H.A. De relatieve beweging van de aarde en den aether, Amsterdam, Zittingsverlag Akad. v. Wet., 1, p. 74-79, (transl.: The relative motion of the earth and the aether) (1892).

32. Fitzgerald G.F. The Ether and the Earth's Atmosphere, Science, v. 13, 390 (1889).

33. Michelson A.A. and Morley E.W. On the Relative Motion of the Earth and the Luminiferous Ether, Am. J. Sc. 34, 333-345 (1887).

34. Miller D.C. The Ether-Drift Experiment and the Determination of the Absolute Motion of the Earth, Rev. Mod. Phys., 5, 203-242 (1933).

35. Pound R.V. and Rebka Jr. G.A. Apparent Weight of Photons, Phys. Rev. Lett., 4(7), 337-341 (1960).

36. Rothall D.P. and Cahill R.T. Dynamical 3-Space: Black Holes in an Expanding Universe, Progress in Physics, 4, 25-31 (2013).

37. Rothall D.P. and Cahill R.T. Dynamical 3-Space: Observing Gravitational Wave Fluctuations and the Shnoll Effect using a Zener Diode Quantum Detector, *Progress in Physics*, 10 (1), 16-18 (2014)

CLIMATE ANOMALIES, THEIR DRIVERS AND TECTONIC CONNECTIONS

GIOVANNI P. GREGORI

*Istituto di Acustica e Sensoristica O. M Corbino (IDASC) – CNR, , via Fosso del Cavaliere 100
00133 Rome, Italy*

BRUCE LEYBOURNE[*]

*Institute for Advanced Studies in Climate Change (IASCC), 12361 East Cornell Ave.
Aurora, CO80014 USA*

Solar influence on climate occurs (by high-frequency modulation) through the "external-way" (Sun → magnetosphere → troposphere) and (by low-frequency) through the "internal-way" (Sun → Earth's endogenous energy → energy exhalation). The percent role of either "way" depends on the spectrum of the solar change. The "external-way" modulation displays substantial uniformity, the "internal-way" a largely variable spacetime anisotropy. A higher percent of "internal" driver implies greater vorticity in fluid Earth, oceans, and atmosphere (the present Sun is very anomalous).

Fluid dynamics of every ionized medium is a very effective and ubiquitous dynamo (according to the Cowling's generalized theorem), from the micro- through the macro-scale. Vorticity supplies altogether increased atmospheric electricity, water condensation and precipitation, larger cloud cover, larger albedo, smaller capture of solar radiation, colder and variable climate.

Very different phenomena are thus correlated to one another. (1) Localized thermal expansion of lithosphere and crust, (2) changes of subsoil electrical conductivity, and (3) gas exhalation, change altogether the electromagnetic (e.m.) coupling between subsoil and atmosphere, thus justifying anomalous and otherwise unexplained events (wildfires), deep-ocean fauna beaching (due to deep water poisoning), and eventually also energy-threshold phenomena (volcanic paroxysm, earthquakes, hurricanes), and seismic/atmospheric (meteorological and/or ionospheric) correlation.

Modern high frequency climate is modulated by solar system orbital physics variation in electromagnetic/gravitational coupling between the Earth and other astrophysical bodies. Solar rotation and orbital influences of the Moon, Jupiter and Saturn have dominant roles in modern climate high frequency modulation. While Earth's longer orbital cycles, such as eccentricity (~100 ka), obliquity (~36 ka), and precession (~26 ka) modulate ice ages, 100 ka year cycles, and various other climate proxies, are known as Milanković cycles.

Ancient palaeo-climate is controlled by (1) Earth e.m. resonance (410 ka years; i.e. eccentricity modulation), (2) Solar System crossing of galactic equatorial plane (~14 Ma), and (3) timing of endogenous energy release (27.4 Ma, Earth's "electrocardiogram"), with different proxies (hotspot volcanic sequences, geodynamic spirals, etc.).

Other cycles are the result of the size and structure of the Earth, with its associated typical timing, such as the so-called "terminations", or the 179 Ma cycle of continent and orogen generation and destruction by weathering and erosion.

1. Introduction - The "external" and "internal way"

Solar-terrestrial relations are a key item in our life. Let us call "climate" the environment which occurs in the space-time domain where life can develop: it is a real very tiny fraction of the whole solid Earth's volume. In a terrella model, 1 : 10,000,000 scale, the Earth is a ball of ~128 cm diameter, while the troposphere, a tiny layer of ~0.01 cm, and the biosphere, including humankind is a lesser mustiness inside it.

The best known solar influence occurs through the solar electromagnetic (e.m.) radiation. A lesser acknowledged - and much more uneven - effect occurs through the corpuscular radiation, i.e. the solar wind, which is identified with a steady expansion of the solar corona. This is not, however, a regular and smooth phenomenon, being rather characterized by inhomogeneity in space and time, as shown by several pictures of the solar corona, observed either during total solar eclipses, or by several space probe images.

The Earth, with its magnetosphere, captures only a fraction (~0.5×10^{-9}) of the spherical expanding

[*] Work supported by Institute for Advanced Studies in Climate Change (IASCC) © all rights reserved.

solar corona. Therefore, the Earth eventually experiences severe consequences from solar wind inhomogeneities. In contrast, during the human time domain solar radiation results, comparably, substantially much more regular, smooth, and stable.

The Earth's climate results therefore controlled by the Sun through both the "external" and "internal way" (figure 1).

The "external way" is a chain of cause-and-effect, beginning from the Sun and its e.m. and corpuscular radiation, through the Earth's magnetosphere, higher atmospheres, and troposphere. This is generally known as "space weather".

The physical mechanism has been extensively and critically discussed in Gregori (2002; shorter accounts in Gregori, 2006a, 2006b, 2009, 2014).

The "internal way" is rather a chain beginning from the solar wind through its long-period e.m. induction inside deep Earth, which modulates the production of endogenous energy, hence of its fluid exhalation into oceans and atmosphere, with a final control on climate.[†]

Climate can be likened therefore to the ensemble of phenomena that occur inside a condenser during its discharge. The upper plate of the condenser - which includes solar wind and higher atmospheres - is powered by an electric current generator (the solar wind). Its lower plate - which is generally underground except on the outer boundary of volcanic plumes – is powered by an electric potential generator (the tide-driven geodynamo).

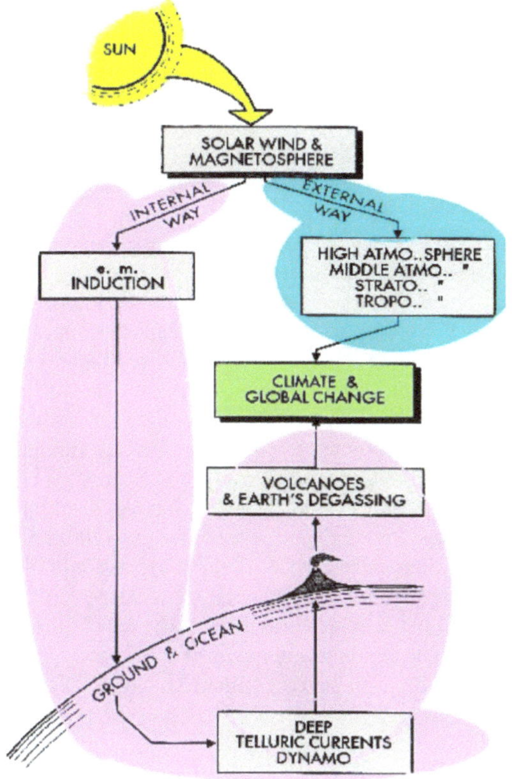

Figure 1 – The "external" and "internal way" in solar-terrestrial relations (after Gregori, 2002).

2. The "Cowling dynamo" and its several manifestations - Atmospheric water condensation, fog and precipitations

Following the previous discovery of the solar magnetic field (Babcock), Larmor proposed (in 1919-1920) an explanation: the violent fluid dynamics inside the Sun or inside every star is associated with very intense electric currents and magnetic fields. The resulting MHD (magneto-hydro-dynamics) phenomena are such that the whole system acts like a very effective dynamo supplied by the endogenous heat.

Extensive modelling is now available, and this explanation results to be very appropriate to explain phenomena inside stars and galaxies.[‡]

[†]Note that e.m. induction in deep Earth cannot occur if the Earth is structured like an "onion" in terms of concentric approximately spherical layers. This is a false paradigm, derived from the need for mathematical simplicity while deriving models for the Earth's interior. Sound physical arguments from college physics show, rather, that real "antennas" underground connect the deep Earth to much shallower layers. Owing to brevity purpose, these items cannot be here repeated.

[‡]The same model was later proposed for the Earth's or planetary interiors. This is, however, physically absurd, due to the lack of any physical source analogous to the thermonuclear energy that forbids total blocking of the system. See Gregori (2002; shorter accounts in Gregori, 2006a, 2006b, 2009).

In 1932 Cowling showed a famous theorem: if a natural system has a *perfect* cylindrical symmetry, no dynamo can be operative.

This caused a real nightmare among solar physicists. An intensive debate was started, essentially still lasting. At present a few tens of proofs of the Cowling theorem are available. Every proof relies on different assumptions. It has been even authoritatively claimed that every proof is biased by some *ad hoc* assumption. But all proofs agree on the same final result. The problem remained apparently unsettled.

For completeness sake, recall that in 1953 Chandrasekhar and Fermi showed the virial theorem for plasmas, and proved that every system of ionized matter with no kind of confinement (such as gravitational) and no other internal source can be self-contained. Rather, it must search to expand in space as much as possible.

Just by a matter of chance, a generalization of the Cowling theorem appeared in Gregori (2002) who did not realize its relevance. No specific assumption is required, rather only the Maxwell's laws. Every system of ionized matter with some internal dynamics - of any origin - must develop a dynamo of either one of two possible topological patterns (figure 2).

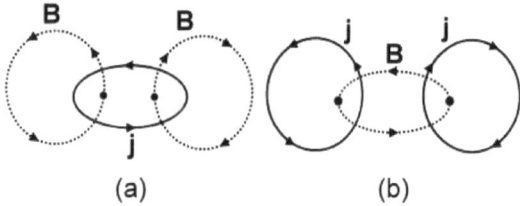

(a)   (b)

Figure 2 – Two topological patterns of the dynamo originated inside a system of ionized matter. The topology of **j** and **E** is the same. After Gregori (2002).

Both these structures lead to an equilibrium state which is a dynamo. But, it can be shown that the case history with toroidal **E** and poloidal **B** (figure 2a) is a state of *unstable* equilibrium, unlike the case of toroidal **B** and poloidal **E** (figure 2b), which is a state of *stable* equilibrium.

That is, only the case of toroidal **B** and poloidal **E** (figure 2b) can be really observed.

However, in the case that the system has a *perfect* cylindrical symmetry, the energy of this stable state is zero, and this gives justice to the original Cowling result.

But, a *perfect* cylindrical symmetry is a simplifying abstraction (much like the concept of point, line, plane etc. in Euclidean geometry) that can find no real equivalence in natural reality. Therefore a dynamo always occurs, being more or less intense depending on

the intensity of MHD motions. It is a ubiquitous phenomenon, on every space-time scale. In the following it will be here briefly called "Cowling dynamo".

It is impossible to report here in detail its several manifestations. Let us give only very few and very brief mentions, from the largest through the smallest space-time domain.[§]

It can explain the filamentary structure of galactic superclusters, or also the unexplained self-collimation of the solar wind, and the **B** of galaxies and stars including their poloidal jets (and the so-called Fermi bubbles of the Milky Way).

On the space-time microscale, tiny convection inside the atmosphere generates a very feeble **E** that displaces the tiny agglomerates of water molecules around every ion. These agglomerates move through the neutral water molecules and collect them operating like "brooms", thus leading to the formation of a tiny droplet. This can eventually float through air (fog or cloud), or precipitate. This mechanism solves – due to the Cowling dynamo - the very well known "mystery" (almost a nightmare for every present climate modeller) of the physical reason for water condensation and precipitation.

Comparably smaller Cowling dynamos coalesce into larger Cowling dynamos, leading to progressively stronger **E** that contribute to the migration of electric charges through the Earth's atmosphere.

Two items need to be considered in some detail: *(i)* a spark or a lightning discharge, and *(ii)* terrestrial gamma flash (*TGF*).

Let us begin and recall a laboratory experiment by Versteegh *et al.* (2008). An electric current is thrown through a tiny water layer located on the point of an electrode. A "light-ball" is generated that lasts a few tens of milliseconds and moves upward for several tens centimetres. Inside it the temperature is a few thousand Celsius degrees, while at its boundary the "ball" is essentially cold. This is a perfect laboratory model for the well-known phenomenon of "ball lightning". This also explains the so-called jellyfish flames (Phillips, 2014b) observed onboard the *International Space Station* (*ISS*).[**] That is, as long as an internal heat source can supply a Cowling dynamo, a toroidal **B** pattern is established, which originates a plasma bottle that confines all hot ions inside it. When the heat source fades off, altogether with the Cowling dynamo, also the "light-ball" rapidly evaporates.

[§]An extensive discussion is given in an 8-volume set in preparation by G. P. Gregori (see a short presentation in Gregori, 2014).
[**]http://science.nasa.gov/science-news/science-at-nasa/2014/10sep_jellyfish/

At present, the physical process is claimed to be essentially unexplained, either of a spark or of a lightning discharge. In contrast, it is triggered whenever a "micro"-Cowling dynamo triggers a former "light-ball". When this fades off, its internal heat triggers a new nearby Cowling dynamo. The process progresses like a domino effect, thus displaying a luminous discharge that moves at some limited speed, thus giving the appearance of a spark or of a lightning.

Huge Cowling dynamos developed inside stormy clouds, and this explains the largest percent of intra-cloud lightning discharges, compared to cloud-ground discharges, etc.

Also several transient luminous effects (*TLE*), which are being progressively observed and reported, can be explained like peculiar similar case histories.

But the huge Cowling dynamos inside clouds have very important additional implications.[††]

Astrophysicists were searching for intense gamma-ray signals from outer space, and they were shocked when they found several intense *TGF* events originated from the lower atmospheric layers, i.e. inside clouds. The most complete account is maybe Marisaldi*et al.* (2014) who report about a systematic regular observations inside the latitudinal belt $\pm2.5°$ carried out by the almost equatorial satellite AGILE. The frequency (for events up to 30 MeV) is ~0.3 per day. But also event of higher energy (up to *100 Mev* and larger) have been reported. A more recent estimate Phillips (2014c) envisages an even dramatically much larger number of *TGF*, i.e. ~1100 per day.

TGF's are known since several years. Their explanation relies on an astute mechanism formerly envisaged by Gurevich in 1961 (although its role was acknowledged only later on). The cross-section of electrons in air rapidly decreases with their energy. There is need for a "seed" by some **E** that causes an initial acceleration of electrons. When they attain a given threshold, the abrupt decrease of their cross-section is such that they can be very rapidly accelerated. Thus they release - by Bremsstrahlung – a high-energy gamma-ray.

A difficulty, however, is about the need for the **E** "seed" (Tavani, private communication, 2014): the Cowling dynamo is the lacking element of the cause-and-effect chain.

This phenomenon also has a very dramatic implication for security: Tavani *et al.* (2013a) have shown, in a multidisciplinary study, that these *TGF*s cause an instant and unrecoverable disruption of every electronic device through a combined action of e.m. signals but also of a neutron flux originated by gamma-rays. This can explain the otherwise mysterious abrupt disappearance of every kind of communication with some aircrafts (either through standard radio links or through portable satellite-connected telephones of passengers). The immediate total loss of control of the aircraft leads to its precipitation like a stone, until its crash at Earth's surface.

3. Solid Earth's effects

Geodynamics, tectonism, chemical geodynamics, seismicity - in its most extreme frequency range, also including the seismic free oscillations [*SFO*], soil exhalation, volcanism, wildfires, deep water fauna beaching (due to water poisoning by floor exhalation), deep water life forms as the beginning of a continuously regenerating food web, etc. - are all items that ought to require a very long discussion.[‡‡]

We want here to mention only the two probably least known items: *(i)* wildfires, and *(ii)* the coupling between underground phenomena including earthquakes, and atmospheric phenomena (both clouds, and ionospheric effects).

NASA's Earth Observatory provides[§§] with a permanently updated planetary mapping of wildfires (a movie with one frame every month since March 2000). Arson events are manifestly a negligible perturbation. The phenomenon has an obvious seasonable dependence, as it requires three ingredients: *(i)* dry underbrush, *(ii)* wind to trigger friction electricity and a micro-spark, *(iii)* soil exhalation of methane and/or inflammable geogas. Wildfires are an effective proxy datum suited to monitor the space-time variation of soil exhalation.

Several features can be recognized in this movie. But the most surprising evidence deals with the Indochina peninsula. Let us briefly describe this interesting and apparently very regular and repetitive phenomenon.

Owing to some evidence that cannot be here reported, the Banda Sea reasonably seems to be an area of anomalous large release of endogenous heat. This has been proposed as a possible trigger of El Niño Southern Oscillation (*ENSO*) driven by gravitational teleconnection (Leybourne and Adams, 1999, 2001) along the Central Pacific Megatrend between the Banda Sea and Easter Island volcanic complexes (Smoot and Leybourne, 2001). An increase of endogenous heat is

---

[††] One possibility, in principle, deals with the exploitation of the electric energy which is inside the atmosphere. But this item is not here discussed.

[‡‡] Refer to the aforementioned 8-volume set in preparation.
[§§] *http://eoimages.gsfc.nasa.gov/images/globalmaps/data/mov/MOD14 A1 M FIRE.mov* retrieved on February 19th, 2014.

associated with a thermal expansion of the mantle beneath the Banda Sea. The local geotumor is uplifted and the Indochina lithosphere slides northward on its slope. Crustal stress propagates through the peninsula, and with it also soil exhalation: wildfires are thus seen to move northward.

When the wildfire moving "wave" reaches (roughly) the Himalaya region, it bifurcates.

One branch moves along the eastern coast of China until it reaches a point, which is seemingly located slightly north of Beijing. This branch ends into some kind of "flash point" of an almost abrupt occurrence of a large number of wildfires.

The other branch apparently disappears while it likely moves (indicatively) along Himalaya, where no dry underbrush can be found. The wildfire "wave" however is likely to proceed until it reaches some area roughly around Karakoram where a "flash point" is observed, even more regular and intense than the aforementioned point north of Beijing.

When this is completed, a new cycle starts from Indochina etc. The phenomenon is impressive and quite regular.

The interpretation is that the lithosphere slides on the slope of the Banda Sea geotumor until it finds some strongly holding obstacles.

One obstacle is located North of Beijing, and, when it yields, it is likely to be the main cause of the violent earthquakes that sometimes hit that region.

The other obstacle is located around the Karakoram area, and it explains the origin of a ~500 km-long region with very high elevation above sea level.[***]

Correlation between wildfire outbreaks in Southern California and Coronal Mass Ejections around Halloween, 31 Oct. 2003, suggest this may have a similar mechanism linked to the Hawaiian hotspot and Guaymas Basin Rift in Gulf of Baja (Leybourne et al., 2004a, 2006).

The correlation between seismicity and peculiar clouds is also a very recent subject of investigation, being rapidly developed. In contrast, the seismic connection with the ionosphere (through e.m. phenomena or airglow) is a much older topic that was systematically

investigated e.g. by the satellite *DEMETER* (by Michel Parrot and co-workers).

4. The Sun's MiniMax and its climatic implications

In the past, on some occasions, when the solar activity has been anomalously low, some tremendous famines eventually occurred, due to the loss of crops originated by cold summer seasons.[†††] This, however, did not occur on the occasion of *every* very low solar minimum.

At present, the Sun is during a maximum that, however, is impressively low and anomalous. Phillips (2014a) proposed the name solar "MiniMax" to denote it (figure 3).

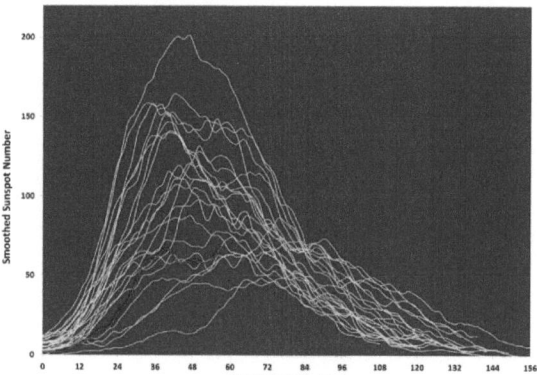

Figure 3 – Several sunspot cycles superposed in order to show the difference between different case histories. This shows the smoothed sunspot number of *Cycle 24* (red), i.e. the present MiniMax, *vs.* the previous *23* cycles since *1755*. After Phillips (2014a).

Note that every sunspot cycle can be very different from all others. But their difference is more evident mostly during the first half of every cycle, i.e. during its first ~5.5 years, while the remaining part appears like a smooth decay towards the next minimum.

Note that sunspot number is a very rough and empirical indicator of solar activity. It relies on an arbitrary weight given to sunspot groups, and on another arbitrary weight given to every solar observatory depending on the kind of instrumentation that it uses, etc.

Several other proxies are certainly better representative of solar activity (e.g. tree rings, or

---

[***]Consider that the time scale for the erosion of continents by weathering can be estimated to be ~179 Ma (Mortari, 2010), which is a comparatively very rapid phenomenon. Therefore, high mountains must be continuously uplifted in order to keep their altitude. The phenomenon depends on a delicate long-time-range balance. The rapidity of mountains weathering has been one of the main concerns – it was a real nightmare - for Leonardo da Vinci: he was correct, but he could not know that mountains are steadily being uplifted.

[†††]Do not confuse, however, these occurrences with the events associated with large volcanic explosions. Compared to the present case histories, those events were a completely different phenomenon. They are the so-called "nuclear-winter scenario". That is, dust in the stratosphere increases Earth's albedo, and climate has less energy supply from solar radiation. Thus it gets colder etc.

isotopic ratios, etc.). But, in general, every proxy is biased by some unknown additional driver (either climatic, or biological, etc.). This item is much debated.

This MiniMax feature is also associated with an anomaly in the extension of sunspots. Figure 4 shows the unusual largest active region seen on the Sun in 24 years that envisages an anomalous heterogeneous spatial large-scale distribution of the solar wind at 1AU.

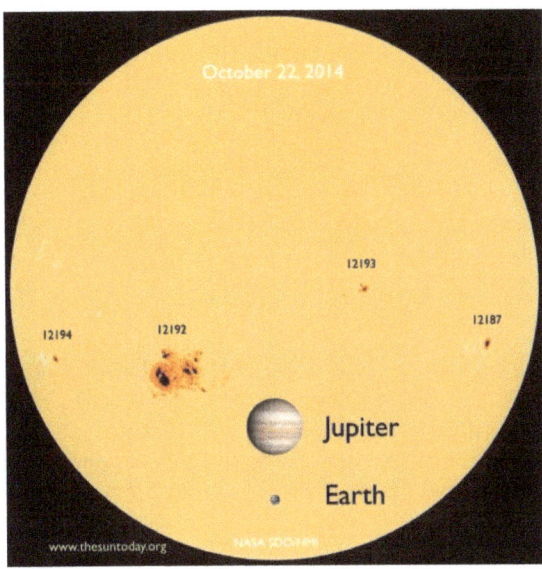

Figure 4 - The Jupiter-sized sunspot AR 12192 is the largest active region seen on the Sun in 24 years. Credit: C. Alex Young/The Sun Today. AR 2192 was actually one of the biggest observed sunspots of all time, ranking 33rd largest of 32,908 active regions since 1874, according to NASA scientists C. Alex Young and Dean Pesnell. After Kramer (2014).

That is, an anomalous large sunspot is likely to be associated with a corresponding overall pattern of the interplanetary environment that is projected outward through space. This ought to be eventually responsible for the generation of anomalous very large storm on the outer planets [such as observed on Saturn and Neptune in 2010, or on Uranus in 2014 (Howell, 2014)]. Outer planets have no seas, hence their "hurricanes" cannot be supplied by sea surface temperature (*SST*) as it occurs on the Earth (Wall, 2011). Therefore, their energy supply must occur through the "internal way", which is modulated by low-frequency e.m. induction by the solar wind.

At present, climate appears to be very anomalous all over the world. A significantly large increase is reported of extreme climatic events from different countries, with fatalities, and no explanation is available.

We propose here a reasonable mechanism, although its final confirmation must wait for careful observations to be collected during the next few decades, in order to check expectation with observed features.

The leading criterion is that solar control on climate is certainly associated with a time varying different percent control, respectively, by the "external way" and by the "internal way".

The "external way" operates by means of phenomena that unavoidably are of planetary extent, and therefore, compared to the effects associated with the "internal way", they result comparably smoother.

In contrast, the "internal way" is associated with phenomena that appear almost "point-like", e.g. like a volcano at Earth's surface. Note, however, that volcanoes are an extreme manifestation of a phenomenon that is ubiquitous and widespread on the planetary scale. Geothermal energy accounts for roughly ~60% of the whole energy balance of the Earth (Gregori, 2002).

Everybody knows that, during the late afternoon of very hot days in summer, an isolated mountain triggers atmospheric convection that finally causes a storm associated with a localized eventually intense shower. Suppose that the percent role of the "internal way" gets eventually larger. In this case, the Earth's surface displays a comparatively larger spatial gradient of endogenous heat release and soil exhalation. That is, it is like in the case that a temporarily large number of "mountains" triggers several convective cells. Let us briefly state that the atmosphere has increased its "vorticity".

Owing to the aforementioned Cowling dynamo process, this triggers an increase of atmospheric precipitation, of atmospheric electrical activity, and a greater development of cloud cover, which implies a greater Earth's albedo with a corresponding decrease of solar radiation captured by the Earth's atmosphere. Climate thus results rainier, colder, stormy, perturbed, and anomalous.

For instance, in Italy a definitely very anomalous 2014 summer occurred. An unusual number of otherwise very rare and exceptional tornadoes were reported. Intense and unpredictable flash-floods occurred with causalities etc. in different parts, and within very few days.

Note that a flash flood can occur depending, first of all, on the specific orographic configuration of the area: when some exceptionally warm and humid air strikes against a mountain range it causes very localized and intense precipitation. But, when no mountain range exists to trigger such an anomaly - such as when dealing with an almost flat landscape - a local area of comparatively larger geothermal flux can sometimes be the ultimate trigger that pushes the system above the threshold for a catastrophe occurrence (this occurred e.g. in October 2014 at the border between Tuscany and Latium).

Other phenomena occurred, such as an anomalous paroxysm of Stromboli (in August 2014 a new boca was opened; the regular tourist excursions had to be stopped). Also around Rome some very unusual new fumaroles appeared. During these same months a renewed concern (maybe) dealt with the Caronia phenomenon, i.e. St. Elmo fires on top of an invisible underground spike that cause abrupt and dramatic spontaneous fires inside some houses. The phenomenon was intense and reported by mass media, during the winter 2003/2004. Recently it apparently strengthened anew.

In addition, occasionally, a few beaching episodes of sperm whales were reported from the Adriatic. These events are absolutely exceptional, and occur only once every several years. Note that, during the aforementioned past occurrence in 2003/2004 of the Caronia phenomenon, exactly during a one week of its most intense and regular manifestation, also water circulation inside the Adriatic Sea reversed, from its standard counter-clock-wise pattern to clock-wise pattern, consistently with an increase of geothermal flow underneath the whole Adriatic sea floor. Multiple earthquake clusters within the Adriatic and Mediterranean Basins in 2003 appear to drive increases in *SST* anomaly magnitudes, geospatial extent, and duration. Manifestations of this phenomenon may be observed in clustered earthquake swarms at the base of the lithosphere, at 10-33 km depths. Burst pulses over several days to weeks appear to precede subsequent *SST* anomalies within days or months after observed seismic swarms. *SST* anomaly patterns overlying earthquake events are hypothesized to be the result of increased heat emission from seafloor volcanic extrusions and/or associated hydrothermal venting (Leybourne et al., 2004b).

Also an anomalous season of tornadoes has been reported from central USA and some unpredictable very intense snow storms in October 2014 on Himalaya (with a few tens of causalities among hikers on Annapurna), etc., while the 2004/2005 U.S. hurricane season was devastating and lightning strikes in the Tampa Bay region the "Lightning Strike Capital" of the U. S. doubled (Leybourne, 2008, 2012).

Summarizing, we guess that, at present, owing to the solar MiniMax, the percent role of the "internal way" is likely to be much larger than during "normal" conditions. This same occurrence happened in the past during some (although not every) period of very low solar minima.

Note that it appears impossible to monitor instrumentally the low-frequency component of solar wind induction into deep Earth. In reality, the unique available "detecting device" seems to be Earth's "climate" according to the aforementioned rationale, or the sunspot area, or the width of outer planets "hurricanes".

5. A short reminder about climatic variability

Just a short reminder can be here given about natural climate variability.

A debated item has been the effect of planetary tides on the Sun, and its feedback on the Earth's climate. The effect is certainly feeble, but real. In particular, it can explain (Tattersall, 2013) some features of paleo-climate, and also some impressive correlation with some long-range time variation of the duration of the length of the day, with a time shift of 30 years, in order to account for friction inside the deep Earth's body that opposes mass displacement.

One can thus eventually recognize inside the climatic data series some effects that appear to be modulated by solar system orbital physics variation in e.m./gravitational coupling between the Earth and other astrophysical bodies. Solar rotation and orbital influences of the Moon, Jupiter and Saturn have dominant roles in modern climate high frequency modulation.

These periodic variations ought to be distinguished from Earth's longer orbital cycles, generally known as Milanković cycles, such as eccentricity (~100 ka), obliquity (~36 ka), and precession (~26 ka), the modulation of ice ages (which is a very complicated and delicate topic that cannot be here discussed in detail), 100 ka year cycles.

In particular, a very remarkable phenomenon deals with the so-called "terminations" evidenced very well in ice cores. Climate gets progressively cooler, until an eventual ice age occurs. Then, it recovers very rapidly. At present, the cause seems to be unexplained. But, this is consistent with the concept of the Earth like a battery (Gregori, 2002) which stores energy. Energy progressively "opens" its way towards Earth's surface. When it affords to open a "channel" through which it can escape above Earth's surface, the discharge capability of the Earth-battery abruptly changes. The battery discharges, the "channel" slowly closes and the recharging process starts anew.

But also other much longer cycles can be recognized e.g. in the magma emplacement rate from the Hawaii hotspot (see e.g. Gregori, 2002 or brief accounts also in Gregori, 2006a, 2006b, 2009, 2014, and references therein). Ancient paleo-climate is controlled by *(i)* Earth e.m. resonance (410 ka i.e. eccentricity modulation), *(ii)* Solar System crossing of galactic equatorial plane (~14 Ma), and *(iii)* timing of endogenous energy release (27.4 Ma, Earth's "electrocardiogram"), with different proxies (hotspot volcanic sequences, geodynamic spirals, etc.).

Recall also the aforementioned 179 Ma cycle observed, among others, in sediments (Mortari, 2010). Consistently with the overall tectonic and geodynamic perspective of the Earth's battery model, this is the likely evidence of the timing of generation of continents and orogens, and subsequent destruction by weathering and erosion.

Atmospheric chemistry suffers by dramatic variations. For instance, the Russian school by Ronov (1982) envisaged the "principle of preservation of life" by which a greater amount of volcanism has always been clearly associated with a greater amount both of carbonate content and of fossil carbon inside the Earth's crust. They claim that this envisages that the greater release of CO_2 by volcanism has been the primary strictly essential "fuel" for the development of the biosphere. That is volcanism is the source for CO_2 and biosphere its sink. Or with no volcanism, no life could exist on the Earth.

But every chemical element, other than carbon, had its long-range cycle. These topics are extremely complicated and the progress is very slow of our understanding, including the crucial role of the biosphere in the control of the whole Earth's system. It is impossible to deal with these items in a short paper.[‡‡‡]

6. Perspective of understanding, natural catastrophe mitigation, "myths" and concreteness

Several present frontiers of research are certainly going to contribute to achieve substantial improvements in our understanding, although the challenge appears very difficult, and it will require a long time.

The ever increasing available capabilities to monitor "climate" from satellite and space platform is steadily improving our observational information that is the real backbone, or "paradigm", of our investigation.

At the same time, we must, however, be extremely "humble" in front of Nature. That is, we must

avoid to impose on Nature with logical schemes and paradigms. Mostly we must refrain from believing that some "simple" model or explanation can get rid of all of what is observed.

In particular, a so-called "numerical model" is an effective tool which is suited to confirm (or not) that our guessed interpretation is correct, and it can also check the amount of detail that we afford to explain by means of a limited set of laws and simplifying assumptions.

But the physical, chemical and biological interpretation of observations is not a numerical model. The model has to be envisaged by means of the exploratory analysis (Tuckey, 1977).

For instance, it makes totally a nonsense to carry out the ongoing - sometimes even very harsh - debate about the role of CO_2 or other gases of anthropic origin that enhance the greenhouse effect, etc.

Important planetary evidence is provided by the recently launched NASA satellite OCO2. Its released dataset, showing the average atmospheric concentration of CO_2 over a period of about 6 weeks late in 2014 (figure 5), indicates that there is CO_2 input in tectonically active oceanic areas. Maxima of CO_2 concentration are observed above the two almost antipodal region of larger release of endogenous energy, roughly located around the Hawaii hotspot and in Botswana. Other maxima occur roughly around the Banda Sea, and in the South American Anomaly. Lesser maxima occur in South Atlantic and in the Indian Ocean (mainly in their respective eastern part), and in the Fiji Island region. Two remarkable maxima are also observed. One is located in north-eastern China that seems not to be associated with anthropic pollution. Rather, it appears to be possibly related to the aforementioned "flash point" of wildfires. The other is located close to Greenland, consistently with several other evidences that suggest that a steady large increase is in progress of endogenous heat underneath the northern polar cap (Phillips, 2014d). This feature seems very likely to be a primary driver in the present ongoing global warming.

[‡‡‡]Also for this refer to the aforementioned 8-volume set in preparation for the discussion of several related items. But no really exhaustive and final discussion can be given.

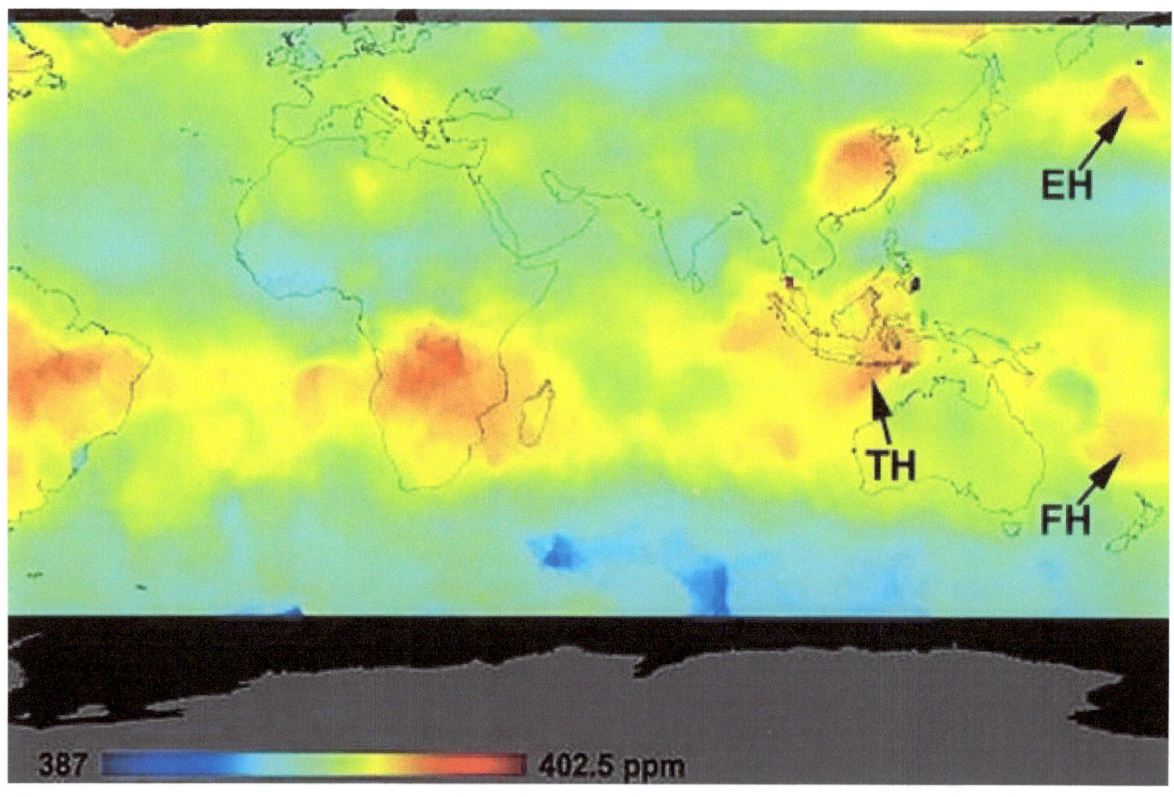

Figure 5 – *"Portions of the initial published OCO₂ data, showing the locations of the three CO₂-hotspots discussed herein. TH=Timor CO₂-hotspot; FH=Fiji CO₂-hotspot; EH=Emperor CO₂-hotspot. They are all apparently associated with tectonically active processes on the underlying seafloor."*[§§§]

It is beyond any doubt that climate is not the same every year, and that a climate change is in progress, just like it happened several times during the entire Earth's history.

It is obvious that humankind has an impact on climate – much like it occurred for every other natural presence inside the Earth's system, of any kind, either living forms or not. Humankind is no exception to this very general and everlasting rule.

It is obvious that humankind must be concerned with avoiding dangerous impacts on "climate", as this can be decisive even for human survival etc.

But, claiming that CO_2 has the main responsibility for climate change is totally unscientific as, according to the present understanding, no serious "proof" in a strict sense can exist that supports such a belief.

Perhaps, the present harshly debated scenario reminds one about an analogous historical circumstance. In the past, on several occasions, a pestilence hit society. For instance, in the XVII century in northern Italy, the dramatic situation was very effectively illustrated in the famous novel "*I Promessi Sposi*" ("*The Betrothed*") by Alessandro Manzoni (1785-1873), which is one of the best known and influential classics of the Italian literature.

The "current opinion" in the XVII century ascribed the main responsibility for pestilence to the "untori" (literally "greasers") who were supposed to disseminate the contagion by some infectious "grease". A real manhunt often occurred. Today, the Italian axiomatic form is still frequently used "caccia all'untore" ("hunt for the greaser") to mean a search for somebody who is simply speculated to be guilty of something with no real proof against him.

[§§§]http://wattsupwiththat.com/2015/01/02/nasas-new-orbiting-carbon-observatory-shows-potential-tectonically-induced-co2-input-from-the-ocean/

In reality, it is now known that sewer rats were responsible for the contagion. But at that time the "official" "generally agreed" science recommended to refrain from taking bath (Sorcinelli, 1998), as water made the pores of the skin widen, thus favouring contagion.[****]

Today, the new "untore" (or "greaser") seems to be CO_2, and this forbids concentration of our research toward prevention from several other forms of anthropic pollution that perhaps, compared to CO_2, are even much more challenging and dangerous.

That is, a sound and realistic humility in front of Nature and of its laws is the fundamental prerequisite in order to manage the very difficult challenge *(i)* of optimizing the interaction between environment and the ever increasing demographic occupation of our planet, and *(ii)* of mitigating the consequences of unavoidable natural catastrophes. Every preconceived "simple" paradigm must be avoided.

References

1. G. P. Gregori, Galaxy – Sun – Earth relations. The origin of the magnetic field and of the endogenous energy of the Earth, with implications for volcanism, geodynamics and climate control, and related items of concern for stars, planets, satellites, and other planetary objects. A discussion in a prologue and two parts. *Beiträge zur Geschichte der Geophysik und Kosmischen Physik*, Band **3**, Heft 3, 471 pp. (2002).

2. G. P. Gregori, The Earth's interior-Myth and science. In "Case studies in physics and geophysics" (ed. by W. Schröder), *Beiträge zur Geschichte der Geophysik und Kosmischen Physik*, special issue (2006/2), Science Editions, AKGG, Bremen-Roennebeck, p. 108-126 (2006a).

3. G. P. Gregori, *NCGT Newslett.*, (38), 34-36 (2006b).

4. G. P. Gregori, 2009. The Earth's interior – Myth and science, *NCGT Newslett.*, (53), 57-75 (2009) [revised version of [2].

5. G. P. Gregori, *NCGT Journal*, **1**, (2), 99-112. (2014).

6. E. Howell, *Space.com*, issued Nov 13, 2014.

7. M. Kramer, *Space.com*, issued Nov 03, 2014.

8. B. A. Leybourne, Satellite Analysis of Joule Heating Mechanism Affect on Climate using GRACE Geoid Interpretations, Earthquakes, Sea Surface Temperature and Lightning Teleconnections, *Italian Embassy Sponsored Climate Workshop*, 25-27 March, 2008, Buenos Aires, Argentina.

9. B. A. Leybourne, Florida hurricanes and grounding of global electric circuits. *34th Int. Geol. Congress*, Brisbane, Australia (2012).

10. B. A. Leybourne, and M.B. Adams, 1999. Modeling mantle dynamics in the Banda Sea triple junction: Exploring a possible link to El Nino Southern Oscillation. *Marine Techn. Soc. Oceans '99 Conf. Proc., Seattle*, Washington, pp. 955-966, Sept., 1999.

11. B. A. Leybourne, and M. B. Adams, 2001. El Niño tectonic modulation in the Pacific Basin. *Marine Techn. Soc. Oceans '01 Conf. Proc.*, Honolulu, Hawaii, Nov. 2001.

12. B. A. Leybourne, A. Haas, B. Orr, N. S. Smoot, I. Bhat, D. Lewis, and G. P. Gregori, *The 8th World Multi-Conf. on Systemics, Cybernetics and Informatics*, (July 18-24), Orlando, FL., pp. 298-299 (2004a).

13. B. A. Leybourne, G. P. Gregori, and C. de Hoop, *NCGT Newslett.*, (38), 3-8 (2006).

14. M. Marisaldi, F. Fuschino, M. Tavani, S. Dietrich, C. Price, M. Galli, C. Pittori, F. Verrecchia, S. Mereghetti, P.W. Cattaneo, S. Colafrancesco, A. Argan, C. Labanti, F. Longo, E. Del Monte, G. Barbiellini, A. Giuliani, A. Bulgarelli, R. Campana, A. Chen, F. Gianotti, P. Giommi, F. Lazzarotto, A. Morselli, M. Rapisarda, A. Rappoldi, M. Trifoglio, A. Trois, and S. Vercellone, *J. Geophys. Res., Space Phys.*, **119**, 1337–1355; doi: 10.1002/2013JA019301 (2014).

15. R. Mortari, Roberto, 2010. *I ritmi segreti dell'universo*, 336 pp., (III ed.) Aracne editrice s.r.l. Roma (2010).

16. T. Phillips, Tony, *Science@NASA*, issued June 10, 2014a.

17. T. Phillips, Tony, *Science@NASA*, issued Sep 10, 2014b.

18. T. Phillips, *Science@NASA*, issued Dec 31, 2014c.

19. T. Phillips, *Science@NASA*, issued Sep 22, 2014d.

20. A. B. Ronov, Aleksander Borisovich, *Int. Geol. Rev.*, **24**, (11/12), 1313-1388. Translated from the Russian edition of 1980 (Izd-voNauka, Moscow, 80 pp.). Reprinted in 1983 by the Am. Geol. Inst. as a monograph (1982).

21. N. C. Smoot, and B.A. Leybourne, *nt.Geol.Rev.*, **43**, (4), (2001).

[****]Hygiene was a real weapon in antiquity. For instance, the ancient Romans searched for great water availability, by means of huge aqueducts. Water was used to manage fire hazard and for hygienic purposes (through several fountains, or the famous public Baths). Thus, large urban settlements could be planned, being reservoirs of soldiers. An efficient road network favoured communication, and a limited number of soldier/policemen permitted to manage a huge territory. This was a real "globalization" of ancient world, by means of an effective communication and exchange between formerly reciprocally very far and isolated regions.

22. P. Sorcinelli, Paolo, 1998. *Storia sociale dell'acqua. Riti e culture.* 194 pp., Bruno Mondadori, Milano.

23. R. Tattersall, *Pattern Recognition Phys.*, **1**, 199-202; doi: 10.5194/prp-1-199 (2013).

24. M. Tavani, A. Argan, A. Paccagnella, A. Pesoli, F. Palma, S. Gerardin, M. Bagatin, A. Trois, P. Picozza, P. Benvenuti, E. Flamini, M. Marisaldi, C. Pittori, and P. Giommi, *Nat. Hazards Earth Syst. Sci.*, **13**, 1127–1133; doi: 10.5194/nhess-13-1127-2013a.

25. J. W. Tuckey, *Exploratory data analysis.* 688 pp., Addison-Wesley Publ. Co., Reading. Massachusetts, etc. (1977).

26. A. Versteegh, K. Behringer, U. Fantz, G. Fußmann, B. Jüttner, and S. Noack, *Plasma Sources Sci. Techn.*, **17**, (2), 1-8; doi: 10.1088/0963-0252/17/2/024014 (2008)..

27. M. Wall, *SPACE.com*, issued on 26 Aug 2011.

CLIMATE OSCILLATIONS (MJO, ENSO & PDO) CONSIDERED WITH VALIDATION OF EARTH ENDOGENOUS ENERGY THEORY

BRUCE LEYBOURNE[†]

Institute for Advanced Studies in Climate Change (IASCC), 12361 East Cornell Ave., Aurora, CO80014 USA

GIOVANNI P. GREGORI

Istituto di Acustica e Sensoristica O. M Corbino (IDASC) – CNR, via Fosso del Cavaliere 100, 00133 Rome, Italy

Climate oscillations with periods ~20 and ~60-years, appear synchronized to Jupiter (12-years) and Saturn (29-years) orbital periods, while Moon's orbital cycle appears synchronized to an Earthy 9.1-year temperature cycle (Scafetta, 2010). The Earth's magnetic moment % decay over the past century reflects 30-year weakening-trends and 30-year strengthening-trends of solar magnetic field, exhibiting the same periods (60-years) well correlated to the Pacific Decadal Oscillation (PDO). In addition, magnetic trends exhibit 3 smaller inflection changes during the 30-year trends which appears correlated: (i) with the 9.1-year Moon orbital cycle and (ii) with El Niño Southern Oscillation (ENSO) patterns overprinted with the 22-year Hale Cycle, more closely associated with a Jupiter affect. Finally the Madden-Julian Oscillation (MJO) 40-day power-spectrum correlates with the period in which Earth experiences a complete solar rotation and interestingly also a north-south oscillation of earthquakes along the Western Pacific rim (Krishnamurti, 2009). A review of the literature indicates that El Niño's have 6-month earthquake precursors (Walker, 1995). In addition the Nation Earthquake information Center database reveals that large 9.0+ earthquakes only occur along the 30-year cooling trend during solar magnetic field strengthening. During this same period Earth also experiences electro-magnetic (e.m.) induction charging, mostly from southern plasma ring-currents coupled to telluric currents in the ridge encircling Antarctica. This transformer effect from the south-pole exerts climate control over the planet via aligned tectonic vortex structures along the Western Pacific rim, electrically connected to the core. This is consistent with the "Earth Endogenous Energy" theory (Gregori, 2002 - Earth as a rechargeable battery/capacitor).

† Work supported by Institute for Advanced Studies in Climate Change (IASCC) © all rights reserved.

Acronyms

AO - Arctic Oscillation
CMB - Core-Mantle Boundary
CMBE's - Core Mantle Boundary Event's
CMSS - Center of Mass of the Solar System
DC - Direct Current
ELOD - Excess Length of Day
e.m. - electro-magnetic
ENSO - El Niño Southern Oscillation
EPR -East Pacific Rise
E.S.I. - electric soldering iron
GOS - Global Oscillation System
GTA - Global Temperature Anomaly
IASCC - Institute for Advanced Studies in Climate Change
LOD – length of the day
MJO - Madden-Julian Oscillation
MPP - Mean Pole Position
NAO - North Atlantic Oscillation
NAVOCEANO – Naval Oceanographic Office
NPO - North Pacific Oscillation
PCI - Pacific Circulation Index
PDO – Pacific Decadal Oscillation
SST - Sea Surface Temperature
SOI - Southern Oscillation Index
TAV - Tropical Atlantic Variability
TD – tide-driven (geodynamo)
WMT – Warm Mud Tectonics

1. Introduction

In considering the validation of Earth Endogenous Energy theory (Gregori, 2002), we are predominately concerned with validation of the internal driver, as the external way as been well documented while the internal way remains a bit more mysterious. Advance studies in climate change are a likely path for uncovering this mystery, as the major frequency modes of climate change seem naturally synchronized with this internal energy and the orbital influences of the more dominant gravitational systems. El Niño Southern Oscillation (ENSO) and Pacific Decadal Oscillation (PDO) climate signatures are in tune with specific 20 and 60 year cycles synchronized to Jupiter (12-years) and Saturn (29-years) orbital periods overlapping lunar cycles synchronized to Earthy 9.1 year temperature

trends and are discussed by Scafetta (2010). The internal way or induction mechanism is naturally linked directly to changes in solar magnetism reflected in strength of the solar winds (Gregori, 2002) and polarity sweeps due to solar rotation driving a much higher frequency (40 day) climate oscillation (Leybourne et. al., 2011) termed the Madden-Julian Oscillation (MJO). Orbital effects from large gravitational bodies not only influence the Earth directly especially during close passes, but induce turbulence within the Sun as the Center of Mass of the Solar System (CMSS) shifts affecting solar tides and interior turbulence linked to solar flaring, sunspots, coronal mass ejections, and solar wind strength. In any case, the ensemble of gravitational perturbations within the Solar System generates, through "collective synchronization of coupled oscillators" (Scafetta, 2010), a resonance response in the Earth system, which is manifested in some oscillation of climate.

1.1. *Internal Way*

The significant changes in solar magnetism have pronounced yet subtle effects to the internal way. One familiar physical manifestation of this affect are the auroral lights emitted from ring currents at both poles as spiraling plasmas decelerate emitting photons. These ring currents at either pole have the ability to induce currents directly into structures within the Earth such as a transformer steps energy down for usage in a local household. Antenna and ring like structures are ubiquitous on planet Earth (Fig. 1: Sea-urchin Model) forming many induction pathways to the core allowing the charging of the planet much like a battery or leaky capacitor (Gregori, 2002). This is the mechanism by which solar wind modulates the performance of the tide-driven (TD) geodynamo, and its associated generation of Earth's endogenous energy. This energy may get dispersed violently as it leaks out creating high magnitude earthquakes, while tidal ebbs and flows gently release Joule heating within volcanoes and hydrothermal vents and can create constant micro-seismicity within active tectonic belts creating what Gregori considers as Warm Mud Tectonics (WMT)

(Gregori and Leybourne, 2014). See explanation in Fig. 7. Remember a higher percent of "internal" driver implies greater vorticity in fluid Earth, oceans, and atmosphere. This increased vorticity relationship is climate change and/or weather depending on the severity and duration according to man's classifications schemes.

2. Tectonic Structures

The tips of the electrical Sea-urchin model, also known as geologic hotspots (Hawaii is a well-known hotspot), takes various forms on planet Earth depending on concentration and power of the sea-urchin spikes (including bunches or clusters of spikes). We have included the term antenna and spike, while other terms include hotspots, rifts, overlapping spreading centers, volcanic complexes/arcs, geo-tumors, super-swells, mega-synclines, triple junctions, micro-plates, and tectonic vortex structures among others. Some confusion is introduced when some terms seem to more define a surface feature, while others seem to include its subsurface extension. Since we are focused on the internal way, the subsurface extension is more important for our discussion; and for our purposes, size does matter mostly for the planetary scale affects we are observing linked to the MJO, ENSO & PDO. For instance planetary scale sea-urchin spikes, electrically connected to the Earth's core at the Core-Mantle-Boundary (CMB), supply energy to Pacific Basin teleconnected oscillation systems outlined in the boxes within Fig. 2.

2.1. *Climate Links*

ENSO frequencies are irregular 3-7 years or more, and seem to be linked to overprinting between the 9.1 year lunar cycles and 11 year sunspot cycles or longer 22 year Hale cycle. The ENSO pressure teleconnection may be explained by mantle density changes due to aforementioned Joule heating between the Banda and Easter Island regions (#3, 4, & 5 in Fig. 2; also see Fig. 3). Density changes can be measured in micro-gals

($u$gals), and the relationship to atmospheric pressure has been quantified by microgravity measurements at approximately 0.33 $u$gals/mbar (Warburton and Goodkind, 1977). Therefore, the ENSO, which drops about 6 mbars during the across-basin pressure shift associated with El Niño, only requires a 2 $u$gals density drop within these structures of the upper mantle to trigger an El Niño. This density change (gravitational teleconnection) can occur rapidly from internal Joule heating linked to tidal affects and caused by electrical emanations from the CMB as a consequence of particularly intense effects of e.m. induction into the Earth, through sea-urchin spikes. It is hypothesized that conductive zones within the upper mantle are associated with electrically stimulated earthquakes at the spinel-perovskite phase change at about 660 km or at the deeper seismic discontinuities at approximately the 840-900 km depth density boundary. The rapid energy transfer below this density boundary between the upper and lower mantle is considered primarily electrical (Leybourne and Smoot, 2005). The ENSO phenomena is also defined by a huge heat signature in the Eastern Pacific. This electrical process is associated with a thermal expansion of the mantle with consequent uplift of superswell or geotumors. The lithosphere slides on their slopes,

favoring overthrust inside megasynclines, according to either Plate Tectonics, Warm Mud Tectonics, or Surge Tectonics. Also see explanation Fig. 7. This originates seismic activity, more or less sparsely.

The PDO is a heat signature in the northern Pacific, considered a temperature teleconnection vs. its counterpart, the North Pacific Oscillation (NPO), an atmospheric pressure teleconnection. They both operate in sync at an approximate 30 year frequency within the Northern hemisphere in a similar fashion to the ENSO. Tectonic modulation is between the Lake Baikal continental rift underlying the atmospheric Siberian High Pressure center and the Aleutian Low Pressure system near the Alaskan trenches (#7 & 8 in Fig. 2).

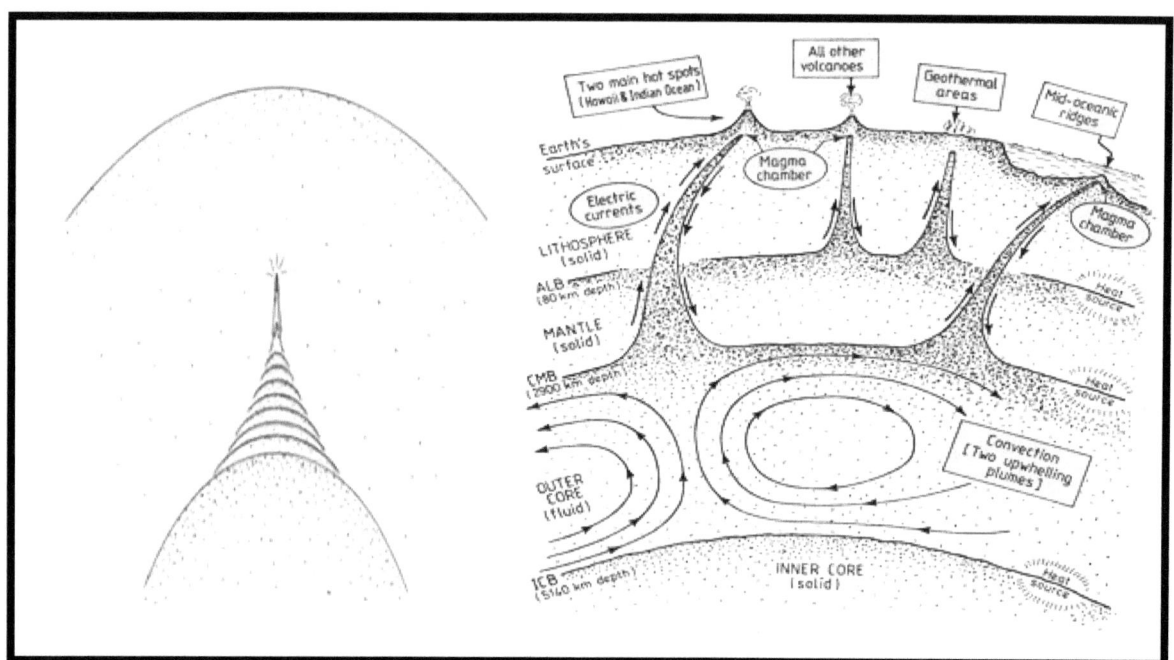

Fig. 1. Earth Endogenous Energy Sea-urchin Spike Model consider a shell of electric currents *j* flowing e.g. at the Core-Mantle-Boundary (CMB). Every given initial minor deviation from perfect spherical symmetry implies a minor bump over which, owing to the Hamilton's principle, the *j*'s must be concentrated. They generate local heating, hence an increase of the local electrical conductivity σ. Hence, owing to Hamilton's principle, the *j*'s must propagate outward, and the process is self-amplifying. The final result resembles an electric soldering iron (E.S.I.) pushed into a block of ice. The cross-section of the upward penetrating spike will shrink over time, and its radial speed will increase (estimated orders of magnitude are from ~ 0 through ~ 20 cm year^{-1}). Several such features can be recognized by means of geomagnetic records (later seen in Fig. 6.) Such a model can be called the sea-urchin model for the Earth's interior as it refers to σ, while the standard "onion" model still holds when referring to the rheological properties of the Earth. After Gregori (1993). For extensive discussion refer to Gregori (2002).

2.2. *Global Oscillations System (GOS)*

The GOS includes: large-scale recurrent teleconnection patterns of variability that influence climate on the regional scale i.e. "large distances typically thousands of kilometers." Among these patterns are the El Niño-Southern Oscillation (ENSO), North Pacific Oscillation (NPO) atmospheric affect vs. Pacific Decadal Oscillation (PDO), Tropical Atlantic Variability (TAV), the North Atlantic Oscillation (NAO), and Arctic Oscillation (AO), all of these can and do largely affect North America and the American monsoon system causing global effects on the rest of the planet as well.

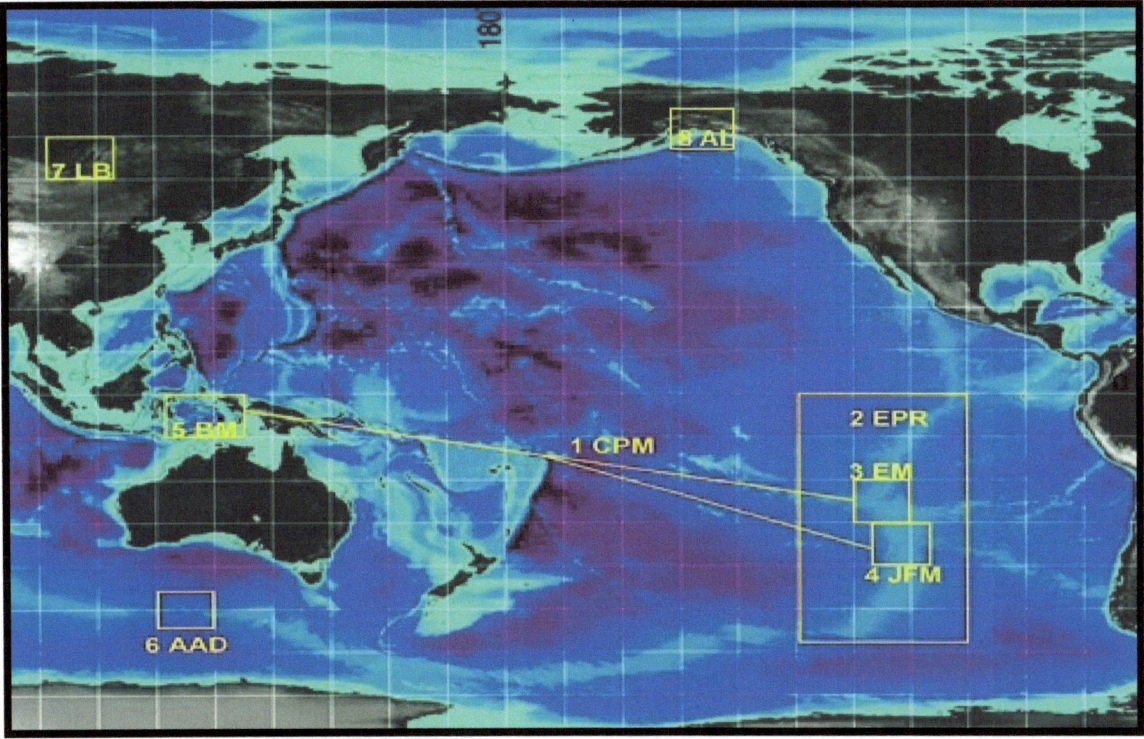

Fig. 2. Bathymetry map depicts geographic locations of teleconnected atmospheric pressure systems in the Pacific Basin for the ENSO and PDO climate oscillations. 1.) Central Pacific Megatrend connects twin rotating micro-plates on the 2.) East Pacific Rise also considered as tectonic vortex structures of the 3.) Easter Island micro-plate and the 4.) Juan Fernandez micro-plate. Connects atmospheric ENSO climate teleconnections via the CPM to the 5.) Banda Sea micro-plate also considered a major Triple Junction in the plate tectonic framework. The 6.) Australian-Antarctic Discordance south of Australia is considered a major sink for south-pole induction energy because of its close juxtaposition and alignment with the south magnetic pole. While the PDO teleconnection runs along the northern Pacific rim connecting 7.) Lake Baikal continental rift high pressure with the 8.) Aleutian low pressure. *Courtesy of U.S. Naval Oceanographic Office (NAVOCEANO): DBDB2-min. gridded bathymetry.*

These major atmospheric pressure teleconnections of the Global Oscillation System all have common tectonic denominators some we've previously discussed. Each system is underlain by highly active tectonic vortex structures, also considered as rotating micro-plates or other planetary scale tectonic feature. Another example is the NAO, an atmospheric pressure oscillation in which the low pressure overlies the Icelandic mantle plume and its teleconnected high pressure center overlies a kink in the mid-Atlantic Ridge along the volcanic complex of the Azores (Leybourne and Smoot, 2005). The Madden-Julian Oscillation (MJO) is a 40 day very high frequency oscillation, similar in observational nature to the ENSO and NPO/PDO climate anomalies. It is closely linked to the Indian monsoon systems. It was realized the same 40-day power-spectrum correlates with the period in which Earth experiences a complete solar rotation. The changing polarity wave emitted from the Sun is hypothesized to control (Leybourne et. al., 2011) the north-south oscillation of earthquakes along the Western Pacific rim uncovered by Krishnamurti (2009) in a 34 year-long United States Geological Survey earthquake database. Examination of this database exhibited nearly 69 swings of these oscillations. These pendulum oscillations in the earthquake frequencies that swing north or south along the Western Pacific Rim and exhibit nearly 10,000 km spatial amplitude of oscillation occurring on an intra-seasonal time scale of 20-60 days with a 40 day power spectrum. Tectonic vortex structures aligned along the Western Pacific Rim affect

the global jet stream patterns or global modes (modalities) of circulation.

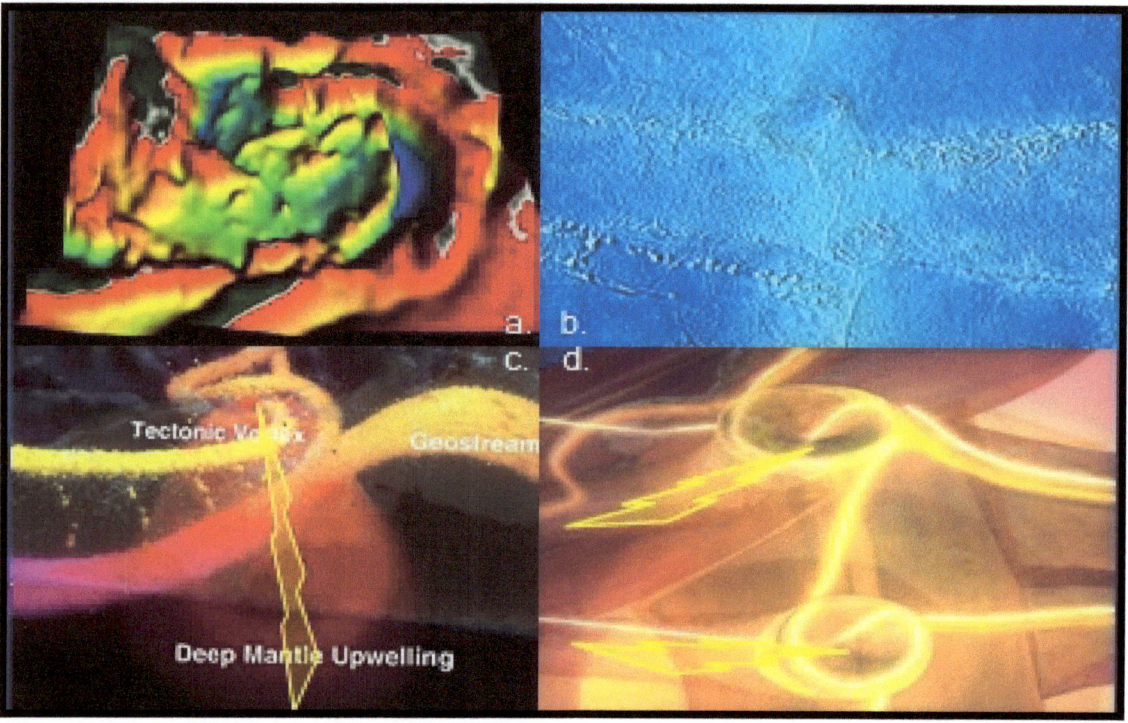

Fig. 3(a). Bathymetric expression of the Banda Sea tectonic vortex; (b). Sea Surface Altimetry (from Sandwell) outlining the Easter Island and Juan Fernandez rotating micro-plates on the East Pacific Rise; (c). Artistic Conceptualization of Banda Sea-urchin Spike driving mantle melt and upwelling diverging convection; (d). Artistic Conceptualization of co-rotating micro-plates driving force from convergent downwelling mantle within twin tectonic vortices along the East Pacific Rise. Lightning bolt image portrays Electric Sea-urchin Spikes. *(Major Shared Resource Center – NAVOCEANO).*

All this sounds heavily related to climate, but still what does this alphabet soup mean? These oscillations systems and related nomenclatures are how climate scientists talk about, understand, and measure the modalities of climate change using indexes of pressure and/or temperature change. In other words each system has its own index of catalogued temperatures and/or pressures related to certain dominant patterns of precipitation events (including lightning), strength and weakness of wind stress fields, position changes of temperature and pressure signals in air and ocean masses related to changes in their flow patterns. These teleconnections are generally associated with anomalies related to each other over large distances and are anomalous to what is considered a normal or average condition, i.e. Reynolds Climatology. And globally

related lightning teleconnections occur through the electric field generated by the "Cowling dynamo" (Gregori and Leybourne, 2014).

For example the January average atmospheric pressure trends from 1949-2001 in the North Pacific related to the NPO exhibit a teleconnection pattern between the Siberian High and Aleutian Low tectonically hypothesized to be modulated by Lake Baikal, a deep continental rift lake, and the Aleutian Trench respectively. Annualized mean atmospheric temperatures, 1949-1998, in the North Pacific linked to Sea Surface Temperature (SST) patterns of the Pacific Decadal Oscillation (PDO) exhibit a teleconnection pattern between Lake Baikal and Hawaii, a well-known geologic hotspot hypothesized to be Joule heated, with episodic heating (Leybourne and Smoot, 2005).

EARTH MAGNETIC MOMENT % DECAY PER CENTURY
(At Earth's Surface)

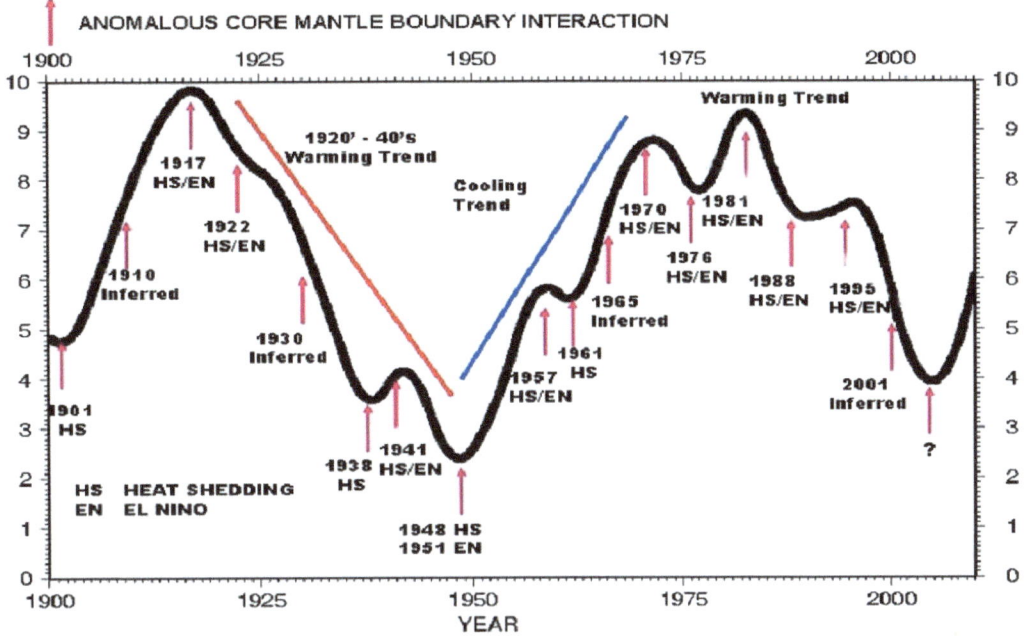

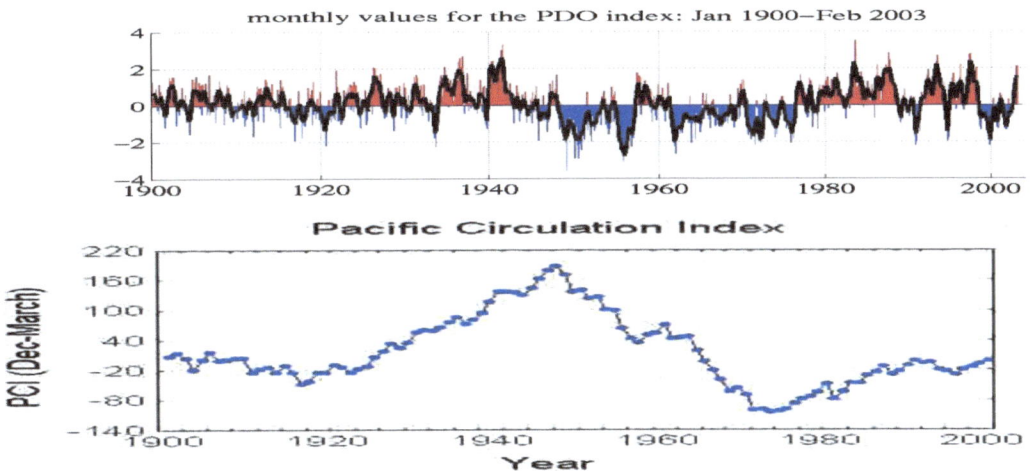

Fig. 4. Scientific Basis for Solar Magnetism Direct Climate Modulation. Major Trends in the Decay Rate of the Earth's Magnetic Moment (Top Fig. - IAGA, 2000 - Models Compiled by Quinn compared to Climatologic Inputs) Long Term Warming 1920-1940 then Cooling in the 1950-1960's until a Second Warming Trend in the 1970's reflect the trends of the Pacific Decal Oscillation (PDO - Middle Fig.). Inflection Points

Correlate as Precursors to Strong El Niño Events, and inversely correlate to trends of the Pacific Circulation Index (PCI), which reflects the wind stress fields near the Aleutians (Bottom Fig).

Magnetic decay data seem to confirm the e.m. mechanism of the Joule heating phenomenon and the modulating affect it has on major climate indexes, such as the Pacific Decadal Oscillation (PDO) and the Pacific Circulation Index (PCI) (Fig. 4).

Thus the Global Oscillation System may be tectonically teleconnected by the processes described as electromagnetic-gravitational teleconnection. This process is hypothesized to have a large influence on atmospheric pressure, modulating jetstream patterns globally. This theoretical construct unifies some of the current theories on tectonics, global electrical circuits, and climate, and when/if confirmed should go a long way in verifying Earth Endogenous Energy theory. Growing observational data is beginning to confirm these relationships. What appear to be remnant Joule "sea-urchin" spikes can be seen in some volcanic modeled mantle magnetic data (Fig. 5). In reality, upon looking at the magnetic anomaly maps of every volcanic area, one can very clearly recognized a "double-eye" pattern - consistently with a planar Direct Current (DC) circuit of the local sea-urchin spikes which supplies the volcano that causes an opposite magnetic field perturbation on the opposite sides of the circuit plane (Gregori, 2002). According to this argument, Fig. 6 can be considered a real map of the location of sea-urchin spikes integrated in time over the whole Earth's history, documented by Earth's surface geomagnetic anomalies. A more artistic idealistic rendering of the concept can be viewed in Fig. 6. The true representation lies somewhere between the noisy data in Fig. 5 and the Fig. 6 fantasy.

2.3. *Earthquakes*

As previously discussed, a north-south oscillation of earthquakes (Krishnamurti, 2009) along the Western Pacific rim with a 40 day power spectrum correlates with the Madden-Julian Oscillation (MJO) and is hypothesized to be the likely driving mechanism associated with changes in polarity of the Sun as Earth experiences solar rotation. Interestingly in addition to ENSO/NPO/PDO relationships with changes solar in magnetism as shown in Fig. 4, ENSO/NPO/PDO also have earthquake relationships. The general rationale of

earthquake links to solar wind is through e.m. induction into deep Earth and control on endogenous heat generation. Telluric current enhancement within the sea-urchin model and the associated Joule heating and consequent thermal expansion of the mantle, gravity sliding of the lithosphere on the slopes of superswells and geotumors, with associated geodynamic seismicity. In this scenario an oversupply of energy can occur both to volcanoes, and to hydro and geothermal heat flow. This eventually increases the probability of occurrence of eruptions, but also a greater injection of energy into oceans and atmosphere, through fluid exhalation across Earth's surface, with consequent implications on climate. This complicate scenario is manifested, and confirmed, by several observed correlations between different indices or morphological features.

Earthquake activity is often associated with magma extrusions, seafloor-spreading events (Walker, 1988, 1995, 1999), increased hydrothermal venting rates and temperatures (Johnson, 2001). Multiple lines of evidence (Quinn, 2010) point to solar magnetism changes and associated inner core jerks tied to triggers of decadal ENSO cycles, and longer centennial 20-30 year global warming/cooling trends such as the Pacific Decadal Oscillation (PDO), which are also linked to wind stress fields of the Pacific Circulation Index (PCI) discussed in Fig. 4. A review of the literature by Daniel Walker retired from University of Hawaii indicates that El Niño's have 6-month earthquake precursors. Walker, as early as 1988, correlates extreme lows in the Southern Oscillation Index (SOI) between 1964 and 1987 with episodes of intense seismicity along the East Pacific Rise (EPR), (Walker, 1988). Swarms of T-phase seismicity are often accompanied by unusually high levels of seismic activity along ridge systems and have been found to coincide with hydrothermal activity and volcanic activity along the Juan de Fuca Ridge (Baker et al 1993; Dziak and Fox, 1993; Embley et al., 1993). Walker states in his rebuttal to Forsyth et al. (1995), "Epicenters and T-phase source locations during intense (El Niño related seismic) episodes have been found to be distributed lengthwise along hundreds of kilometers of ridge systems and laterally displaced from ridge crests by tens of kilometers" (Walker, 1988;

Walker and Hammond, 1990). It seems tectonic expansion episodes along ridges (geostreams) related to El Niño have already been well documented. The irregular periodicity of El Niños defined by a low index phase of the SOI, makes the

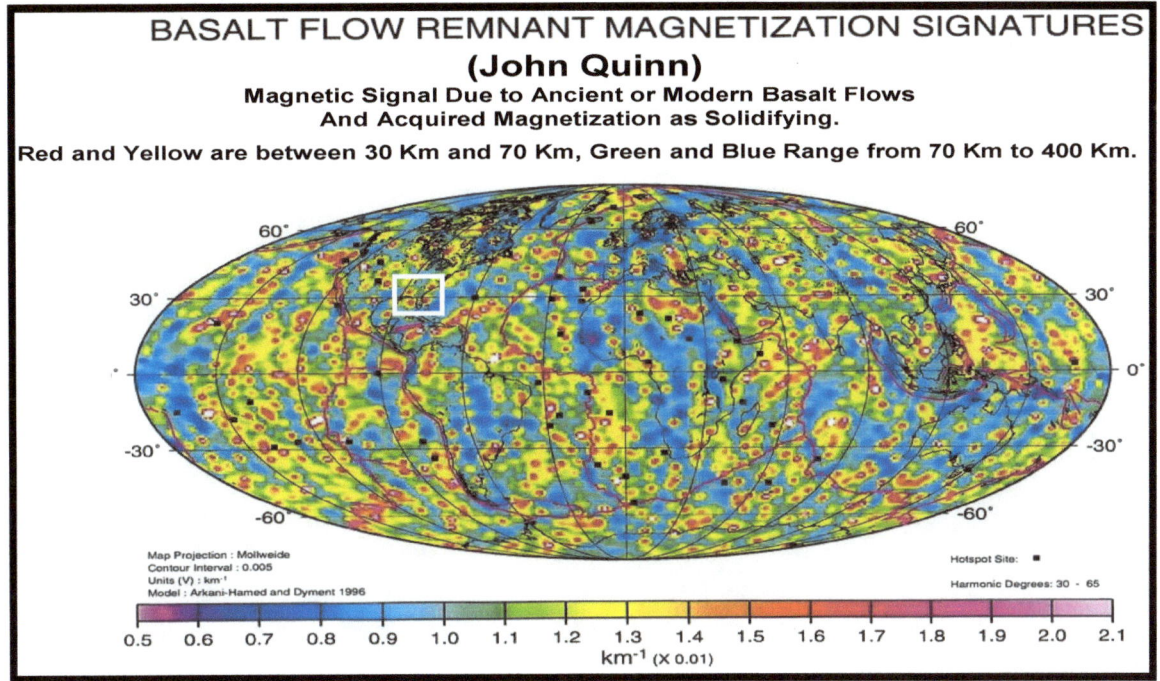

Fig. 5. Quinn's Remnant Magnetization Signatures are considered the tips of Gregori's "Sea-urchin Spikes" within the Endogenous Earth Energy theoretical framework. Some may also be meteorite impacts. White square indicates ancient features in Southeast U.S. likely associated with Triassic Rifting.

matching irregularities of seismicity very hard to dismiss as merely coincidence (Walker, 1995; Forsyth et al., 1995). The causative link of direct and indirect thermal effects associated with massive submarine volcanic and hydrothermal activity accompanying these seismic swarms as Walker speculates, is not the only link of El Niños and seismicity. SST anomalies within the vicinity of large clustered burst of earthquakes at the base of the lithosphere just months after the earthquakes are to be in some cases believed to be associated with locations of hydrothermal venting during the earthquakes, caused by endogenous heat associated either with Joule heating or with friction heat originated by overthrust of lithospheric slabs.

In addition the Nation Earthquake Information Center database (Table 1) reveals large 9.0+ earthquakes only occur along the 30-year cooling trend during solar magnetic field strengthening (Fig. 4). So, why were there no Richter 9.0+ earthquakes between, 1964 and

2004? The large 9.0 earthquakes have started back up with the Dec. 2004 Boxing Day Sumatra earthquake and tsunami, after a 35 year hiatus. Some large earthquakes did strike in Kobe and San Francisco for example, although still devastating they were not > 9.0's. The table also shows several large > 8.0 earthquakes in remote areas that did not affect large population centers. We are hypothesizing Richter 9.0+'s did not strike during the weakening solar magnetic field trend because the planet was discharging and could not build up enough charge to deliver the 9.0+. But we believe the 9.0+ earthquakes will continue striking during the current charging interval until the next solar magnetic weakening trend begins in ~2035.

Note, however, that in general a greater amount of crustal stress does not necessarily imply an immediate earthquake, which is rather the result of an accumulation of stress until it reaches a rupture threshold. However, since the aforementioned correlation deals with a 30-year time span, this accumulation effect might be, perhaps, shorter than 30-years.

3. Discussion

As discussed above, the Nation Earthquake Information Center database reveals that large 9.0+ earthquakes (Table 1) only occur along the 30-year cooling trend

during solar magnetic field strengthening. It is hypothesized that during this same period Earth also experiences e.m. induction charging, mostly from southern plasma ring-currents (Fig. 7) coupled to telluric currents in the ridge encircling Antarctica (Fig. 8). This transformer effect from the south-pole exerts climate control over the planet via aligned tectonic vortex

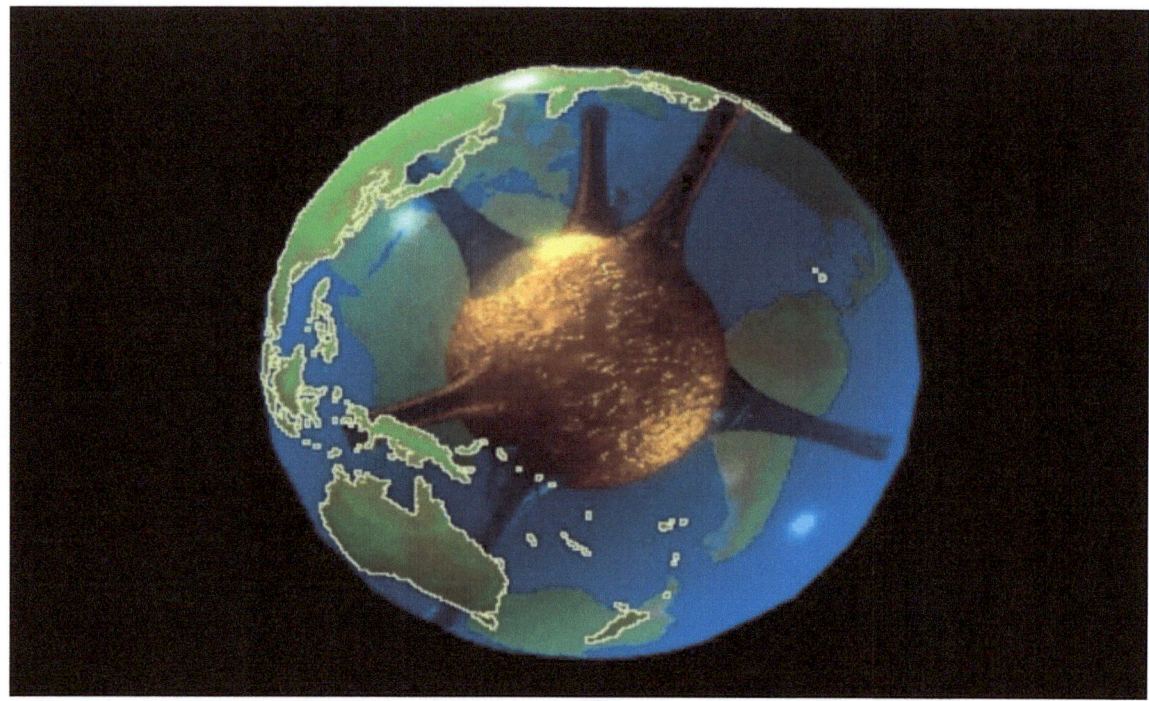

Fig. 6. "Sea-urchin Spikes" Artistic Rendition. *(Major Shared Resource Center – NAVOCEANO).* This is an artistic and greatly oversimplified representation of natural reality. Indeed, the sea-urchin spikes are to be considered a very frequent occurrence, more or less grouped into bunches that eventually coalesce vs. time into larger spikes. Wherever a dense amount occurs of spikes, the thermally expanded mantle uplifts a superswell. The lithosphere slides on its slopes, and this is the primary cause of every form of geodynamics (this model is briefly denoted as "warm mud tectonics", *WMT*). In the case, however, that also a return flow of lithospheric material occurs underground, being a continuation of the lithospheric material that slides at Earth's surface, the geodynamic model is called "surge tectonics". In particular a huge bunch of sea-urchin spikes ought to be located in order to justify the so-called Hawaii hotspot.

structures along the Western Pacific rim (See Fig. 2), electrically connected to the core primarily at the south pole near the Australian Antarctic Discordance (Leybourne, 2014) along a downwelling tectonic vortex on the Southeast Indian Ridge just south of Australia (Leybourne and Adams, 2000). These ideas are consistent with Gregori's "Earth Endogenous Energy" theory. Thus explaining Earth as a rechargeable battery

or leaky capacitor, and begins to necessitate a full paradigm shift in the understanding of climate science.

Some large earthquake events occurred in the mid 1960's and smaller earthquakes increased frequency just before the Great Pacific Climate Change in 1970's which raised the Pacific Basins average temperature 1°C. Geomagnetic jerks are associated with these increased earthquake periods and can be observed in various ways, including seismic anisotropy methods

(Woodhouse and Dziewonski, 1984). In addition, other proxies such as Excess Length of Day (ELOD) and Mean Pole Position (MPP) are used. The fact that the magnetic pole movements are affected indicates that the origin of these jerks is in the Earth's core (Quinn, 2010).

Geomagnetic jerks have been the subject of study for several decades (Courtillot and Le Mouel, 1984; Alexandrescu et al. 1995, 1996; Bellanger et al. 2001, 2002; Chambudot and Mandea, 2005). Externally, chaotic tidal forces associated with the relative motions

(i.e., Barycentre motion) of the Earth, Moon, Sun planets and galactic influences (Gregori, 2002) may trigger geomagnetic jerks. Barycentre motion applies mechanical torque to Earth. Solar-terrestrial interactions between the Sun's magnetic field, the Earth's magnetic field and the solar wind can also act as an external trigger (Ducruix et al., 1980). These events are associated with and can also be referred to as Core Mantle Boundary Event's (CMBE's) as noted by (Woodhouse and Dziewonski, 1984).

Table 1. Hiatus of >9.0 earthquakes during ~30 year cycle linked to discharging and decreasing solar magnetic field strength trend of the Earth, while reactivation of >9.0 earthquakes which switch on during ~30 year cycle linked to charging during the increasing trend of solar magnetic field strength. (Leybourne - non published observation). http://earthquake.usgs.gov/earthquakes/eqarchives/year/mag8/magnitude8_1900_date.php

Date – UTC – Time		Latitude	Longitude	Magnitude	Fatalities	Region
7/9/1905 9:40		49	99	8.4		Mongolia
1/31/1906 15:36		1	-81.5	8.8	1000	Colombia-Ecuador
11/11/1922 4:32		-28.553	-70.755	8.5		Chile-Argentina Border
2/3/1923 16:01	Hiatus?	54	161	8.5		Kamchatka
2/1/1938 19:04		-5.05	131.62	8.5		Banda Sea
8/15/1950 14:09		28.5	96.5	8.6	1526	Assam-Tibet
11/4/1952 16:58		52.76	160.06	9		Kamchatka, Russia
3/9/1957 14:22		51.56	-175.39	8.6		Andreanof Islands, Alaska
5/22/1960 19:11		-38.29	-73.05	9.5	1655	Chile
10/13/1963 5:17		44.9	149.6	8.5		Kuril Islands
3/28/1964 3:36		61.02	-147.65	9.2	125	Prince William Sound, Alaska
2/4/1965 5:01	Hiatus	51.21	-178.5	8.7		Rat Islands, Alaska
12/26/2004 0:58		3.295	95.982	9.1	227898	off the west coast of northern Sumatra
3/28/2005 16:09		2.074	97.013	8.6	1313	Northern Sumatra, Indonesia
9/12/2007 11:10		-4.438	101.367	8.5	25	Southern Sumatra, Indonesia

2/27/2010 6:34		-35.846	-72.719	8.8	577	Offshore Maule, Chile
3/11/2011 5:46		38.322	142.369	9	28050	Near the East Coast of Honshu, Japan
4/11/2012 8:38		2.311	93.063	8.6		off the west coast of northern Sumatra

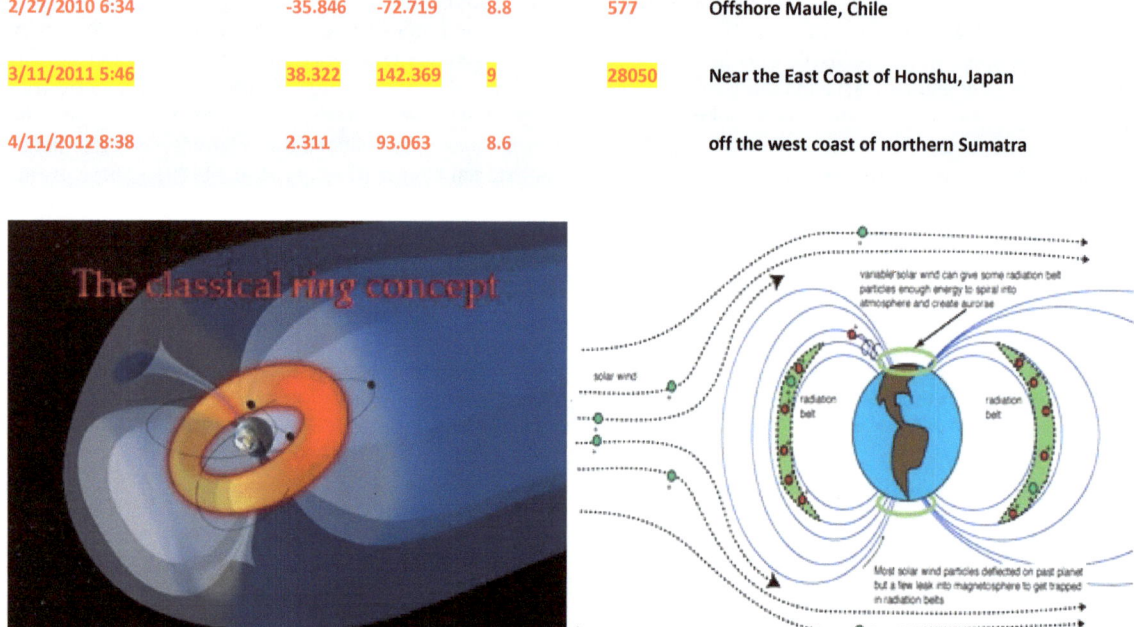

Fig. 7(a). Classical Ring Concept with Radiation Belt and Vortex tied to the Solar Wind. (b). Cross-section of the Radiation Belt with Polar Plasma Rings fed trapped particles from the Radiation Belts and Solar Winds pushing the Polar Plasma Ring speeds. Southern Polar Plasma Ring Currents are coupled to Tectonic Ridge Structure Ringing Antarctic in Fig 8. Sea-urchin Spikes along the Tectonic Ridge Ring complete the connection to the Earth's core (images - NASA).

Fig. 8. Polar stereographic projection of the Southern Ocean region with structural features, such as fracture zones and mountain ranges portrays the ridge encircling Antarctica outlined in black (Smoot, in press).

The entire jerk process may take less than a year to as much as five years. A periodic Joule-heating transfers

may occur instantaneously from the Core-Mantle-Boundary to the base of the lithosphere, (lightning from below driving earthquakes). Geophysical and environmental parameters like the Global Temperature Anomaly (GTA- as reflected by the PDO) and MPP and ELOD often exhibit turning points or inflection points (also see inflection points in Fig. 4) at jerk events. Some events are global in nature, while others are regional. As a rule, the dipole portion of the geomagnetic field is associated with global jerks, whereas the non-dipole portion of the geomagnetic field is associated with regional jerks (Quinn and Leybourne, 2010). This general rule is consistent with the aforementioned general rationale, as a comparably very deep effect originated by e.m. induction from the solar wind has planetary implications, hence it affects the dipole field. In contrast, comparably less deep and more local effects are associated with regional patterns, hence with non-dipolar magnetic field components. Exceptions do occur; however geomagnetic jerking events can be isolated by computing dipole moments and non-dipole moments using a series of geomagnetic field models dating from 1900 to present separated at 5-year intervals, but interpolated to 1-year intervals (Fig. 4), in order to match the annual means data of the other

4. Conclusions

In conclusion, investigating the geophysical, electrical, gravitational, and magnetic teleconnection signals of large scale tectonic features (Fig. 2) may unravel local teleconnection to climate influences such as the monsoons MJO, ENSO and NPO/PDO climate change anomalies.

Consider Earth endogenous energy generation using the 40 day solar rotation earthquake driver hypothesis. North-south oscillation of earthquakes with a 40 day power spectrum are suspected to drive the MJO (Leybourne et. al., 2011). While Hale cycle Sun spot correlations may drive clustered earthquake events triggering large Sea Surface Temperature anomalies resulting in El Niño (Leybourne et al., 2005).

In this respect, two crucial drivers are to be considered: *(i)* the non-neutral electric charge of the solar wind, which implies a non-neutral corpuscular local precipitation of particle above the ionosphere (the ionosphere has a very low electrical conductivity, and electric charge re-distribution requires some long time lag), and *(ii)* the role of the "Cowling dynamo" with energy supplied by thermal convection in the atmosphere (which is controlled by endogenous energy exhalation into the atmosphere) (Gregori and Leybourne, 2014).

All aforementioned concepts also apply to the longer term Milankovitch Orbital Cycles. Explaining relationships between major climate change and the 3 main orbital frequencies of eccentricity, obliquity, and precession. The overall explanation opens a unification conceptual framework.

References

1. Alexandrescu, M.., and D. Gibert, G. Hulot, J-L Le Mouel, and G. Saracco; Detection of Geomagnetic Jerks Using Wavelet Analysis, *Journal of Geophysial Research*, **100**, B7, pp. 12557 - 12572 (1995).

2. Alexandrescu, M, and D. Gibert, G. Hulot, J-L Le Mouel, and G. Saracco; Worldwide Wavelet Analysis of Geomagnetic Jerks, *Journal of Geophysical Research*, **101**, B10, pp. 21975 - 21994 (1996).

3. Baker, E.T., G.J. Massoth, G.A. Cannon, R.A. Freely, J.E. Lupton, R.E. Thomson, J.F. Gendron, B.J. Burd, and R.W. Embley, 1993. Temporal and spatial patterns of chronic and event hydrothermal plumes at the CoAxial Vent Field, Juan de Fuca Ridge, July 1 to October 20, 1993. *EOS. Trans. AGU,* 74:619.

4. Bellanger, E., J. –L. Le Mouel, M. Mandea, and S. Labrosse, Chandler Wobble and GeomagneticJerks, *Physics of Earth and Planetary Interiors*, v. **124**, pp. 95 – 103 (2001).

5. Bellanger, E., D. Gilbert, J.-L. Le Mouel; A Geomagnetic Triggering of Chandler Wobble Phase Jumps? *Geophysical Research Letters*, **29**, pp. 28/1 – 28/4 (2002).

    Chambodut, A. and M. Mandea; Evidence for Geomagnetic Jerks in Comprehensive Models, *Earth Planets Space*, **57** (2): 139 - 149 (2005).

6. Courtillot, V. and J. -L. Le Mouel; Geomagnetic Secular Variation Impulses, *Nature*, **311**:709-716. (1984).

7. Ducruix, J., C. Courtillot, J. -L. Le Mouel; The Late1960's Secular Variation Impulse, theEleven Year Magnetic Variation, and the Electrical Conductivity of the DeepMantle, *Geophysical Journal of the Royal Astronomical Society*, **61**: 73 – 94(1980).

8. Dziak, R.P., and C.G. Fox, 1993. Seismo-acoustic evidence of a dike injection along the CoAxial Segment, Juan de Fuca Ridge. *EOS. Trans. AGU,* 74:619.

9. Embley, R.W., W.W. Catwalk, I.R. Jonasson, S. Peterson, D. Butterfield, V. Tunnicliffe, and K. Juniper, 1993. Geologic interferences from a response to the first remotely detected eruption on the Mid-Ocean Ridge: CoAxial Segment, Juan de Fuca Ridge. *EOS. Trans. AGU,* 74:619.

10. Forsyth, D.S., D.S. Scheirer, and K. C. Macdonald, 1995. Link between El Ninos and seismicity is still missing. *EOS Trans. AGU.* 76 (17):175.

parameters studied. This technique isolates only the strongest jerks (Quinn, 2010).

11. Johnson, H.P., R.P. Dziak, C. R. Fisher, C.G. Fox, and M.J. Pruis, Earthquakes' Impact on hydrothermal systems may be far-reaching, *EOS Trans. AGU,* **82** (21)*,* 233-236, 2001.

12. Krishnamurti, T. N., Krishnamurti, R, Sagadevan, A. D., Chakraborty, A., Dewar, W. K., Clayson, C. A., and Tull, J. F., 2009, Space-time Structures of Earthquakes, *Meteorology and Atmospheric Physics*, 2009, v. 105 (no. 1-2), p. 69-83.

13. Gregori, G.P. and Leybourne, B.A., Climate anomalies, their drivers and tectonic connections, *21st Annual Natural Philosophy Alliance Conference Proc.*, Baltimore MD, Nov. 2014.

14. Gregori, G.P., Galaxy-Sun-Earth Relations: The origins of the magnetic field and of the endogenous energy of the Earth, *Science Edition, Arbeitskreis GeschichteGeophysik,* ISSN: 1615-2824, W. Schroder, Germany, 2002.

15. Gregori, G.P., Geo-electromagnetism and geodynamics: Coronal discharge from volcanic and geothermal areas, *Physics of Earth and Planetary Interiors,* **77**, 39-63, 1993.

16. Leybourne, B.A, N.C. Smoot, and B. Longuis, World Encircling Tectonic Vortex Street – Geostreams Revisited: The Southern Ring Current EM Plasma-Tectonic Coupling in the Western Pacific Rim, *EGU 2014 Conference*, April 2014, Vienna, Austria.

17. Leybourne, B.A., M.I. Bhat, S. Mishra, Solar Polar Rotation Driving Madden-Julian Oscillation's Seismic Teleconnections, *New Concepts in Global Tectonics International Conference on Earth Dynamics Proceedings – Perceptions and Deadlocks,* Kanyakumari, India, Sept. 2011, p. 56.

18. Leybourne, B.A., G.P. Gregori and C.F. deHoop, Gulf of California Electrical Hot-Spot Hypothesis: Climate and Wildfire Teleconnections, New Concepts in Global Tectonics Newsletter, no. 38, 2005.

19. Leybourne, B. A. and N. C. Smoot, 2005.Tectonic Links to the Global Oscillation System: A Unification Concept of Climate Modulation by Internal Joule Heating Teleconnections, *AGU Chapman Conf. on Tropical-Extratropical Climate Teleconnections*, Hawaii.

20. Leybourne, B.A., and M.B. Adams, "The Australian Antarctic Discordance: Pressurized vs. Non-pressurized Ridge System", *MTS Oceans 2000 conference in Providence*, RI, Sept. 2000.

21. Quinn, J.M. and B.A. Leybourne, Jerks as Guiding Influences on the Global Environment: Effects on the Solid Earth, Its Angular Momentum and Lithospheric Plate Motions, the Atmosphere, Weather, and Climate. *American Geophysical Union San Francisco Annual Fall Meeting*, 13-17 Dec. 2010 (A33A-0152).

22. Quinn, J. M., 2010, Global Warming: Geophysical Counterpoints to the Enhanced Greenhouse Theory, Dorrance Publishing Co., Inc. Pittsburgh, PA. September 1, 2010. 118 pgs, ISBN: 978 - 1 - 4349 - 0581 – 9.

23. Scafetta, N., 2010, Empirical evidence for a celestial origin of the climate oscillations and its implications, *Journal of Atmospheric and Solar-Terrestrial Physics*72 (2010) 951–970.

24. Walker, D.A., Seismicity of the East Pacific: correlations with the Southern Oscillation Index? *EOS Trans. AGU.* **69**, 857 1988.

25. Walker, D.A., and Hammond, S.R., Spatial and temporal distributions of T-phase source locations on the Juan de Fuca and Gorda Ridges. *EOS Trans. AGU.* **71**, 1601. 1990.

26. Walker, D.A., More evidence indicates link between El Niños and seismicity. *EOS Trans. AGU.* **76** (33), 1995.

27. Walker, D.A., Seismic Predictors of El Niño Revisted. *EOS Trans. AGU,* **80** (25), 1999.

28. Warburton, R.J., J.M. Goodkind, The influence of barometric-pressure variations on gravity.*Geophys. J.R. Astr. Soc.* 48:281-292, 1977.

29. Woodhouse, J.H., and A.M. Dziewonski, 1984. Mapping the upper-mantle: Three-dimensional modeling of earth structure by inversion of seismic wave forms, *J. Geophys. Res.* 89. 5983-5986.

APPLICATION OF SRT TO
CALCULATING MERCURY'S PERIHELION ADVANCE

JAMES KEELE

3313 Camino Cielo Vista, Santa Fe NM 87507

e-mail jkeele9@cybermesa.com

This article shows how SRT as applied to charged particles may be applied in an analogous manner to gravitational bodies. With such application and the creation of a computer program, the perihelion advance of Mercury is calculated. The results are compared with results from calculations based on General Relativity Theory. Important characteristics of gravity are illuminated, including gravity laws.

1. Introduction

Two earlier GED articles, by T.E. Phipps, Jr. [1} and M.E. Hassani [2], have inspired this author to work on the analogy between electromagnetic and gravitational forces. This author has shown in [3] how the concepts of SRT apply to interactions between relatively moving charged particles. Here he extends those concepts to gravitational interactions between relatively moving heavenly bodies.

Based on an analogy to the relationship between relatively moving charges, Phipps [1] developed formulas for velocity–dependent gravitational potentials. That article inspired this author to derive the velocity dependent gravitational force using SRT. This derivation begins by presenting a one-to-one correlation of the forces of relatively moving heavenly bodies to the forces between the relatively moving charges. This correlation depends on the hypothesis that a gravitational field exists about a massive body and has characteristics similar those of an electric field about a charged particle.

Linear four-dimensional Minkowski space-time is standard for SRT. This space-time is easier to visualize than the non-linear curved space-time of GRT. Understanding the complexities of formulating the math of GRT and of solving the resulting differential equations is not necessary to comprehend the present article. However, a formula derived from SRT is employed, and its derivation involves the complexity of four-vector math. This derivation is not presented in this paper, but is referenced [4].

The following analysis uses the mathematical sign convention that the attractive force between masses is positive, and that both masses as applied in the equations have positive values.

2. Analogies

Traditional force magnitude and the Coulomb Law for elecric force magnitude is:

$$F = Gm_1 m_2 / r^2 \Leftrightarrow k q_1 q_2 / r^2 \quad , \qquad (1)$$

where G is the gravitational constant, m_1 and m_2 are the attracting mThe correspondence between the Newtonian law for gravitaasses, and r is the distance between the masses, and where $k = 1/4\pi\varepsilon_0$, with ε_0 being the permittivity of free space, q_1 and q_2 are the charges of the interacting particles, and r is the distance between the particles.

The three dimensional electric field $\mathbf{e}_c$ pervading the space about a charged particle is defined below, and is analogous to the three dimensional gravity field $\mathbf{e}_g$:

$$\mathbf{e}_c = kq\mathbf{r}/r^3 \Leftrightarrow \mathbf{e}_g = Gm\mathbf{r}/r^3 \qquad (2)$$

where $\mathbf{r}$ is a position vector extending from the charge or mass and r is its magnitude.

The forces $\mathbf{F}_e, \mathbf{F}_g$ are analogous, and defined as:

$$\mathbf{F}_c = q_1 \mathbf{e}_c \Leftrightarrow \mathbf{F}_g = m_1 \mathbf{e}_g \quad . \qquad (3)$$

Application of the SRT version of the electric field of the moving charge as seen by a stationary charge is:

$$\mathbf{e}_c = kq\mathbf{r} \Big/ \gamma^2 r^3 \left[1 - (v^2/c^2)\sin^2\theta \right]^{3/2} \qquad (4)$$

where v is the magnitude of $\mathbf{v}$, the relative velocity between the two charges, $\gamma = 1/\sqrt{1 - v^2/c^2}$, θ is the angle between $\mathbf{r}$ and $\mathbf{v}$. The constant $k = 1 / 4\pi\varepsilon_0$, and c is the speed of light.

Place the test charge q_1 at the stationary point, and you can create an expression that represents the total electrodynamics force between the stationary charge and the moving charge. This force consists of the electric Coulomb force and the magnetic force. This expression is good for relative velocity from zero up to c. This is why this author calls this formula "The Basic Electromagnetic Law". Relativists might say Eq. (4) determines only the electric field, and the magnetic field is something else. But the magnetic force emerges as the difference in the electric field of a stationary charge and the electric field of the moving charge as observed by a stationary charge or observer. The derivation of this SR formula can be found in books on Relativity such as Dr. Wolfgang Rindler's book [4].

If the hypothesis that the gravity field has characteristics similar to the electric field is correct, then the gravity field of a moving heavenly body as viewed from a stationary heavenly body (or stationary observer) may be expressed as:

$$\mathbf{e}_{\mathbf{g2}} = G\gamma m_2 \mathbf{r} \Big/ \gamma^2 r^3 \left[1 - (v^2 / c^2)\sin^2\theta \right]^{3/2} \quad . \qquad (5)$$

The terms in (5) have the same meaning as the terms defined for (4), except G, gravitational constant, and m_2, the mass of the moving body, replace k and q_2, respectively. Multiplying (5) by m_1 creates a formula for the gravity force between relatively moving heavenly bodies with relative velocities from zero to c

Eq. (5) is where the complete analogy between moving charges and moving masses breaks down. Notice the inclusion of the gamma factor, γ, in the numerator of (5). This expresses the relativistic mass increase of the moving mass m_2. Charges are invariant on being transformed from one inertial frame to another, and an isolated moving charge does not need a gamma factor.

A study of (5) reveals the intensity of the gravity field increases when the relative velocity vector $\mathbf{v}$ is at a right angle to the $\mathbf{r}$ vector. When the velocity vector $\mathbf{v}$ is in line or parallel to the $\mathbf{r}$ vector, the intensity of the gravity field is reduced depending on the magnitude of the velocity. This author makes no claim that this application of SRT theory replaces GRT, but does make the claim this force relationship is valid, and can be used to study relationships between relatively moving heavenly bodies. As evidence for the power of this application of SRT, this author has created a computer program that calculates the Mercury perihelion advance from a formula derived from (5). The program is also used for calculating the perihelion advance of Venus and Earth.

3. The Formula for Force Between Relatively Moving Heavenly Bodies ($v \ll c$)

Upon canceling gamma factors in the numerator and denominator of (5), and including the mass m_1 to create force per (3), and applying the binomial series to the factors in the resulting denominator, and eliminating higher orders of v^2 / c^2, one arrives at the following formula:

$$\mathbf{F_{12}} = \frac{Gm_1 m_2 \mathbf{r}}{r^3} \left[1 + \frac{v^2}{c^2}\left(1 - \tfrac{3}{2}\cos^2\theta \right) \right] \qquad (6)$$

Eq. (6) is the formula that can replace Newton's Law for gravity for most applications. Note that this force law has a term dependent on speed v and direction angle θ, defined above for Eq. (4). This dependence on speed and direction is a consequence of SRT, and are included in the velocity dependent formula for gravity. Some other authors [1,2] who write about the velocity-dependent gravity field do not include the direction dependence.

Eq. (6) is analogous to the formula that represents the force between a stationary charge and a current element. That electrodynamics formula can be viewed in this author's paper [3].

4. A Computer Program for Testing the Gravity Law (6)

This author generated a Pascal program employing the speed and angle dependent term of (6) for computation on a standard IBM type PC. The intent was to calculate the perihelion advance of the planets Mercury, Venus, and Earth. This was satisfactorily achieved. While the program will not be presented in detail in this article, the following paragraphs will present the essential elements of the program. The program offers insight into how the perihelion advances of elliptical orbits occur. Double data type is employed in the program to provide 15-16 significant figures in the computations.

4.1. *The Basic Approach of the Computer Program*

A planet will traverse a perfect elliptical orbit about the Sun if only the Newton Gravity Force Law applies. This law varies in strength as $1/r^2$. But if the Relativistic Gravity Force Law applies, then the planetary orbit varies from a true elliptical orbit by a very small amount. For Mercury, the ratio of the added force caused by relativity to a true Newton force is 3.86×10^{-8} at perihelion. Therefore, the velocity of the planet used for the relativity term can be determined from the true elliptical orbit with little error. Also, the small planetary drift created by the relativity term may be modeled separately from the larger Newton Gravity Force.

An elliptical orbit is mathematically defined [5] for a given planet using the ellipse center reference origin and x and y coordinates. The ellipse semi-major axis, a_{smajor}, coincides with the x-axis and the semi-minor axis, b_{sminor}, coincides with the y-axis. The focus of the ellipse, the center of rotation of the Sun and Mercury, is placed on the positive x-axis at a distance equal to ε, the eccentricity. Perihelion is then on the positive x-axis at distance of a_{smajor} from the center of the ellipse. Inputs [6] to the program are the particulars of a given planet such as its mass, m_2; semi-major axis, a_{smajor}; eccentricity, ε; and speed at perihelion, v_p; and the time required for one complete orbit, T. Other inputs include the gravitational constant, G and the mass of the Sun, m_1, and the speed of light c.

The mass of the planet is modeled as moving for one orbit counterclockwise around the circumference of the rigidly defined orbit, starting at perihelion. Calculations are done with x values determined at regular intervals on the x-axis ($\Delta x = a_{\text{smajor}} / 20$). Computations are done for one-half of the orbit; then this same routine (for the first half of the orbit) is rotated 180 degrees and used again to calculate values for the second half of the orbit. Appropriate values generated in the first half of the orbit are carried over to the second half of the orbit. For the second half orbit, the focus is moved to the negative x-axis a distance of ε.

The total orbital path is divided into 80 spatial intervals corresponding to equal lengths Δx along the x axis, and 80 corresponding intervals of time, Δt with various numerical values computed by the program. Also computed at each cumulative x value are the radial distance from the sun to the planet, the angle of $\mathbf{r}$ with respect to the x-axis, the angle of the planet's velocity vector with respect to the x-axis, the angle of the planet's velocity vector with respect to the radial vector $\mathbf{r}$, the angle between successive $\mathbf{r}$'s, and the speed of the planet.

The speed of the planet is calculated based on the difference in potential energy from one interval to the next giving up that energy to a difference in kinetic energy. The initial potential energy and kinetic energies of the planet are calculated at perihelion.

The speed and angle-dependent force contribution from (6) is:

$$\Delta \mathbf{f_{12}} = \frac{Gm_1 m_2 \mathbf{r}}{r^3} \frac{v^2}{c^2} \left(1 - \frac{3}{2} \cos^2 \theta \right) \qquad (7)$$

This force implies a radial acceleration contribution of value:

$$\Delta a = \Delta f_{12} / m_2 = \frac{Gm_1}{r^2} \frac{v^2}{c^2} \left(1 - \frac{3}{2} \cos^2 \theta \right) \qquad (8)$$

The angle θ in (8) is the angle between the radial vector $\mathbf{r}$ and the velocity vector $\mathbf{v}$ of the planet's motion in its orbit. This angle is generally near $\pi/2$, making the term involving θ small. For the planet Mercury, the factor in parentheses varies from a low of .937 to 1.0.

The program starts at perigee, where the velocity of the planet in the radial direction is zero. It numerically integrates (8) to find a velocity correction, and numerically integrates that to find a radial displacement correction.

The displacement increments from all 80 iterations for one complete orbit are summed and called s_{total} and d_{total}. These totals are at right angles with respect to each other. So they combine to:

$$s_{\text{hypotenuse}} = \sqrt{(s_{\text{total}})^2 + (d_{\text{total}})^2} \qquad (9)$$

To calculate the perihelion advance one needs knowledge of how ds changes with respect to the angle change $d\theta$ of a radial vector $\mathbf{r}$. This is accomplished by noting that the circumference of a circle extends for an angle of 2π, so:

$$\Delta\theta = s_{\text{hypotenuse}} / r \quad . \qquad (10)$$

A suitable r is obtained for the elliptical orbit by averaging the major and minor semi-axis radii: $r = (a_{\text{smajor}} + b_{\text{sminor}})/2$. The accumulated ds is then divided by this r at the end of the iterations. The perihelion advance is computed for 100 Earth years.

4.2. Computed Results for the Perihelion Advances of Three Planets

Table 1 compares results computed using the force formula (7) and the results computed using GRT. We see perihelion advances expressed in arc seconds per 100 Earth years for three planets.

Table 1. Comparison of SRT computed results with GRT calculated results.

Planets:	Computed Advance	GRT Calculated*	%Difference
Mercury	42.60	42.98	−0.89
Venus	8.18	8.62	−5.1
Earth	3.63	3.84	−5.5

*Data supplied by Wikipedia (internet):

en.wikipedia.org/ wki/ Test_of_general_relativioty

The program for Mercury perihelion advance was run without the term containing θ in Eq. (7). The results were 43.92 arc-sec per 100 Earth years, an increased difference of 2.18% compared to the GRT value.

5. Discussion of Results

The GRT calculated results presented in Table 1 are in agreement with observed results referenced to the ICRF, International Celestial Reference Frame. The computed advance results are relative to the Sun or more correctly relative to the center of rotation of the Sun and planet. As such they are essentially referenced to the ICRF. The computed results could not be in such close agreement with the observed results unless the force formula (7) is valid as a term of Eq. (6). The same computer program was used for calculating the perihelion for the three planets with only the input parameters changed. Eq. (7) causes the program to track the various perihelion advances of the three planets. The close agreements of the advances predicted by Eq. (7) strongly suggest that Eq. (6) is a valid refinement to Newton's gravity law.

The results unequivocally support the v^2/c^2 term in (6). It is to be noted that the v^2/c^2 term has a factor depending on the angle between the position vector **r** and the velocity vector **v**. It represents the increase of the gravity field of the moving body when the velocity vector of the moving body is at or near perpendicular to the position vector between the two bodies. It shows a decrease in the gravity field of the moving body when it is moving away. The angle dependent term, having θ as a parameter, was derived from SRT and is 'locked' with the v^2/c^2 term. The results for Mercury computed with the θ term show closer agreement with the GRT values than the results without that term. The difference is not large enough to form a definite conclusion, based on the computation alone, about the θ term.

Close agreement with the GRT calculated value was achieved with this computer model for the Mercury perihelion advance (-0.89% difference). This suggests that the computer model is a good one if the GRT value and observed value for Mercury are reliable. When the model is applied to the perihelion advances of Venus and Earth, there were larger error differences of 5.1% and 5.5% respectively. These error differences amount to an averaged difference of -0.34 arc sec/hundred Earth years for the computer model as compared to the GRT values.

The gravity law Eq. (6) could replace Newton's gravity law for some applications. It is strongly supported by the computed perihelion advances when compared with the observed advances. Here is a quote from Einstein's paper of 1915 when he discloses his application of GRT to the Mercury perihelion advance: "This calculation leads to the planet Mercury to move its perihelion forward by 43" per century, while the astronomers give 45"±5", an exceptional difference between observation and Newtonian theory. This has great significance as full agreement."[7]

As criticism to this SRT-based approach to gravity, here is a quote from Dr. Wolfgang Rindler, "Several attempts have been made to construct also new theories of gravitation within special relativity, but this can only be done at the heavy cost of abandoning the equivalence principle or a 'natural' interpretation of SR." [4] p. 76. He also said, referring to the linear approximations to GRT, "We end this chapter with a brief discussion of a subject that is important in many practical applications of GR, from gravitational waves to the physics of black holes: the linear approximation. This approximation to GR is usually much simpler to apply than GRT itself, though it must be applied with care; it sometimes gives results which in no way approximate to those of the full theory." [4] p. 188. This author admits to not having a full understanding of GRT. He speculates that GRT is a non-linear version of the linear SRT.

6. Expanding Universe

Eq. (5) extended to include the second mass is the more universal gravitational law:

$$\mathbf{F_{12}} = Gm_1m_2\mathbf{r} \Big/ \gamma r^3 \left[1 - (v^2/c^2)\sin^2(\theta)\right]^{3/2} \quad (11)$$

Eq. (11) is good for the relationship between two relatively moving masses with relative speed from 0 to c.

Observe from (11) that, if the direction of the relative velocity is in the direction of the $\mathbf{r}$ vector (moving away from each other), then the $\sin^2(\theta)$ term goes to zero and the equation is left with one γ factor in the denominator. Therefore, as the relative velocity increases to values large compared to c, the force of gravity attraction is considerably reduced, not only by large r, but also by large v. This formula applies to the expanding Universe, and is the more general of the two gravity formulas since (6) is restricted to relative speed much less than c. Most relative speeds encountered are much less than c.

Eq. (11) predicts that a spinning disc whose plane is in the vertical direction should weigh slightly less than when it is spinning at the same rate in a plane in the horizontal direction.

7. Conclusion

The perihelion advance of Mercury was one of the first 'proofs' utilized to support GRT. Now the same argument can be made with the same logic to support SRT. It is very significant that the perihelion advance of Mercury can be calculated from SRT. The use of Eq. (6) could simplify calculations that are complicated in GRT. SRT may model cosmic considerations when the distance is equal to or greater than the distance between the Sun and Mercury. That includes most of the universe. And space may be interpreted with four dimensions instead of 'curved space'.

These results suggest that the gravity force acts like a field force, similar to an electric force field. So a *gravity field* force is the probable cause of gravity. Thinking of gravity as a field versus as being 'curved space' is valid for the calculation of the Mercury Perihelion Advance. The gravity laws expressed as Eqs. (6) and (11) are supported with the results of the computer computations presented in this paper.

Acknowledgment

I wish to acknowledge Dr. Cynthia Whitney for helping me edit this paper and for making suggestions for improvements.

References

1. T.E. Phipps, Jr., "More on Gerber's Velocity-Dependent Gravitational Potential", Galilean Electrodynamics, **22**, 4, p. 68 (2011).
2. M.E. Hassani, "Combined Gravitational Action – Second Part", Galilean Electrodynamics, Special Issue, **22**, SI 3, p. 43. (2011).
3. J. Keele, "SR Theory of Electrodynamics for Relatively Moving Charges", Proceedings of the Natural Philosophy Alliance, 16th Annual Conference of the NPA, 25-29 May, 2009, Vol. 6, No. 1, p. 118.
4. W. Rindler, **Essential Relativity**, Revised Second Edition, p.101 (Springer-Verlag, New York, Heidelberg, Berlin, (1977) pgs. 100-103.
5. J.J Tuma, PhD, **Engineering Mathematics Handbook,** 2nd Edition, McGraw-Hill Book Company, p. 44 (1979).
6. CRC, **Handbook of Chemistry and Physics,** 52nd Edition (1971-1972), pgs. F-145-150
7. Einstein's Paper: "Explanation of the Perihelion Motion of Mercury from General Relativity Theory",Anatoli Andrei Vankov, IPPE, Obninsk, Russia; Bethany College, KS, USA; anatolivankov@hotmail.com.

THE DEVELOPMENT OF GENERAL RELATIVITY, REPULSIVE GRAVITATION AND THE THEORIES IN ASTROPHYSIC--the achievements and errors of Einstein in relativity--*

C. Y. LO

Applied and Pure Research Institute
15 Walnut Hill Rd., Amherst, NH 03031

Einstein's covariance principle is proven to be invalid in physics. Due to inadequate understanding of the principle of causality, errors were not recognized. Owing to inadequacy in mathematics, many distort Einstein's equivalence principle and created additional errors. Einstein's theory of measurement adapted in the Riemannian geometry is proven invalid. However, the valid measurement is just what Einstein used in calculating the bending of the light. Thus, the interpretation of Hubble's law as due to the Doppler effect is questionable. Although accurate predictions were made from the static equation, for the dynamic case the Einstein equation is invalid since there are no bounded dynamic solutions. Thus, the 1993 Nobel Prize press release on gravitation is incorrect. Nevertheless, some erroneously claimed that they have found bounded dynamic solutions for the Einstein equation; and D. Christodoulou was even elected to be a member of U.S. National Academy of Sciences (2012) for his errors against Gullstrand. It is found that the existence of dynamic solutions requires additionally a gravitational energy-stress tensor with the anti-gravity coupling. To have the gravity related to an electromagnetic wave, it is also necessary to have additionally an energy-stress tensor of photons with an antigravity coupling. This necessary existence of the anti-gravity coupling for the dynamic case implies that the energy conditions in space-time singularity theorems of Hawking and Penrose cannot be satisfied. Thus, their theorems are actually irrelevant to physics. The invalid assumption of a unique sign for all couplings was also used by Yau, Schoen, and Witten to prove the positive mass (energy) theorem. Nevertheless, the Fields Medal considered their misleading theorems as achievements twice (1982 & 1990). The existence of anti-gravity coupling also leads to the investigation of $E = mc^2$. It is concluded that the electromagnetic energy is not equivalent to mass, and the charge-mass interaction, which is repulsive for the static case, is discovered. Moreover, the charge-mass interaction necessarily established the unification of electromagnetism and gravitation. This would explain the weight reduction of charged capacitors, the pioneer anomaly discovered by NASA, etc. Since the charge-mass interaction exists in both new and known physical processes, the unification would raise a new revolution in physics that may explain many new and long standing puzzles in physics. For instance, the charge-mass interaction is not included in quantum mechanics. Also, the existence of repulsive gravitation implies that the theoretical foundation for the notion of black holes, which is based on gravity is always attractive, requires to be justified again.

Key words: principle of causality; dynamic solution; charge-mass interaction; current-mass interaction; $E = mc^2$.
04.20.-q, 04.20.Cv
* Supported by the Chan Foundation

1. Introduction

Since Einstein's prediction on the bending of light ray was verified, general relativity is dominating the theories of astrophysics [1]. Subsequently, predictions based on the linearization of Einstein equation were verified [1, 2] although some invalid claims were also masqueraded as the truth. [1) Nevertheless, any prediction related to the dynamic case of the non-linear Einstein equation, such as the speculations of the existence of black holes and the expansion of the universe [2), has not been confirmed [3]. It was popular to believe that general relativity has superseded Newtonian gravity, but this belief was due to careless mathematical errors [3]. Moreover, currently it was believed that general relativity is not valid for microscopic phenomena [2], and no prediction of general relativity has been verified in any laboratory on earth, and there was no prediction made on non-massive sources [1, 2]. In this paper, we shall address these problems that stand in the way of theoretical developments.

Einstein's general relativity was based on two principles, the equivalence principle, the covariance principle, and the Einstein equation [4, 5]. However, Einstein's theory is actually not self-consistent. For instance, counter examples have been found for the covariance principle [6]. Many theorists and Einstein failed to understand that the Einstein equation with massive sources is valid for only the static and stable cases [7-9] and that different sources can generate very different gravities [10]. In fact, some sources can generate gravity that is repulsive to a mass [10, 11] although only attractive gravity is generated by the mass in Newtonian theory. (However, because of Einstein and his followers over confidence of Einstein's claims, these were over-looked.) It is on such a basis Einstein's conjecture of unification between gravitation and electromagnetism can be established [12]. Einstein and Pauli [13] did not understand this need of new interactions and thus they failed to establish Einstein's conjecture of unification.

The unification of gravitation and electromagnetism would cover many new areas as well as old areas in physics. For instance, it would include the weight reduction of a charged capacitor that was a puzzle rejected by many theorists because of their bias and their failure to explain such phenomena. In fact, it is potentially a new revolution in physics to solve new problems and many long standing issues. In this paper, we shall discuss these problems related to general relativity.

2. The Equivalence Principle of Einstein and its Misinterpretations

In 1911 Einstein started his theory of gravitation by assuming the equivalence of a uniformly accelerated system K' and a stationary system K with a uniform Newtonian gravitational potential ϕ. Many assume the Newtonian metric form,

$$d\tau^2 = (1 + 2\phi)\, dt^2 - dx^2 - dy^2 - dz^2, \qquad (1)$$

that Fock [14] has proved to be impossible from the equivalence principle. (According to this, surprisingly Fock claimed and the Wheeler School[3] agreed that Einstein's equivalence principle is invalid.) From metric (1), Einstein derived the correct gravitational redshifts, but an incorrect light velocity that leads to only one half of the observed light bending angle [4].

However, Einstein's equivalence principle of 1916 assumes the equivalence of a uniformly accelerated system K' and a stationary system of coordinate K with an *unspecified* metric form that generates a uniform gravitation [4]. Einstein made clear also that the uniform gravitational field is generated from a space-time metric, but is not a Newtonian potential. Moreover, concurrent with Einstein's equivalence principle of 1916, Einstein makes the claim of the Einstein-Minkowski condition as a consequence [4, 5].

However, in the press release of the 1993 Nobel Committee [15], the equivalence principle was claimed as the identity between gravitational and inertial mass (due to Galileo and Newton), but not as Einstein's equivalence principle although it has been confirmed by experiments (see eq. [3'd]). [4] However, since Einstein did not provide an example to illustrate his principle, a reader could mistake it as the 1911 assumption of equivalence. It is not until 2007 that a metric for uniform gravity [16] was published as follows:

$$ds^2 = (c^2 - 2U)\, dt'^2 - (1 - 2U/c^2)^{-1} dx'^2 - (dy'^2 + dz'^2), \qquad (2)$$

where $c^2/2 > U(x', t') = (at)^2/2$, "a" is the acceleration of system K'(x' y' z') with respect to K(x, y, z, t) in the x-direction. Metric (2) shows the time dilation and space contractions clearly. Here, dt' is defined locally by cdt' = cdt − (at/c)dx'[1 − (at/c)^2]^{-1}. Moreover, metric (2) is equivalent to the metric, $ds^2 = (c^2 - a^2t^2)dt^2 - 2at\, dtdx' - dx'^2 - (dy'^2 + dz'^2)$, that was derived by Tolman [17]. It was a surprise that U is actually time dependent, and this explains the earlier failed derivation of such a metric [18]. Nevertheless, due to the popular "Einstein's elevator" of Bergmann [19], Einstein was often falsely accused of ignoring the tidal force by Thorne [20]. [5]

To illustrate the equivalence principle further, consider a disk K' uniformly rotating w. r. t. an inertial system (x, y, z, t), a metric for the disk of space K' (x', y', z') is derived [21]. According to Landau & Lifshitz [22], the metric is

$$ds^2 = (c^2 - \Omega^2 r^2)dt^2 - 2\Omega r^2\, d\phi'dt - dr^2 - r^2 d\phi'^2 - dz'^2 \qquad (3)$$

where Ω is an angular velocity relative to an inertial system K (x, y, z, t), z and z' coincide with the rotating axis, and $r^2 = x^2 + y^2 = x'^2 + y'^2$. Metric (3) is equivalent to its canonical form,

$$ds^2 = (c^2 - \Omega^2 r'^2)dt'^2 - dr'^2 - (1 - \Omega^2 r'^2/c^2)^{-1} r'^2 d\phi'^2 - dz^2 \qquad (3'a)$$
where
$$cdt' = cdt - (r\Omega/c)rd\phi'[1 - (r\Omega/c)^2]^{-1}. \qquad (3'b)$$

However, (3'b) is not integrable [21] because local time dt' is related to different inertial systems at different r or time t. Thus, to obtain the correct space contractions, one must first transform the metric to a canonical form such that the space contractions are clear.

The fact that the local time t' is not a global time was a problem that leads to the rejection by the editors of the Royal Society [21]. This rejection is incorrect since validity of metric (3') can be derived theoretically with special relativity. Experimentally, the time dilation from metric (3'a) for the local metric, $ds^2 = c^2 dT^2 - dX^2 - dY^2 - dz^2$, is

$$dT = [1 - (r\Omega/c)^2]^{1/2}\, dt'. \qquad (3'c)$$

From (3'b) the local clock resting at K', if observed from K, would have

$$dt' = dt. \quad \text{and} \quad dT = [1 - (r\Omega/c)^2]^{1/2}\, dt. \qquad (3'd)$$

Moreover, as Kundig [23] has shown, the time dilation (3'd) is valid for a local clock fixed at K'[6].

Hence, Einstein's equivalence principle has experimental supports although his claim [4] on this time dilation was invalid [21]. Thus, the 1993 Nobel Committee press release should not frivolously reject this principle; especially since it was done implicitly [15].

A source of confusion is that many have mistaken Pauli's invalid version [24] as Einstein's equivalence principle although Einstein has made clear it is a misinterpretation [25]. For instance, in the book "Gravitation" [1] of Misner, Thorne and Wheeler, there is no reference to Einstein's equivalence principle (i. e. [4] and [5]). Instead, they misleadingly refer to Einstein's invalid 1911 assumption [26] and Pauli's invalid version [24]. This is due to that they do not understand the related mathematical theorems.

Unfortunately, many universities, research institutes such as Harvard, MIT, Princeton and Stanford, as well as the 1993 Nobel Committee are victims of such a distortion. The mathematical theorems [27] related to Einstein's equivalence principle are as follows:

Theorem 1. Given any point P in any Lorentz manifold (whose metric signature is the same as a Minkowski space) there always exist coordinate systems (x^μ) in which $\partial g_{\mu\nu}/\partial x^\lambda = 0$ at P.

Theorem 2. Given any time-like geodesic curve Γ there always exists a coordinate system (the so-called Fermi coordinates) (x^μ) in which $\partial g_{\mu\nu}/\partial x^\lambda = 0$ along Γ.

In these theorems, the local space of a particle is locally constant, but not necessarily Minkowski.

However, after some algebra, a local Minkowski metric exists at any given point and along any time-like geodesic curve Γ. In a uniformly accelerated frame, the local space in a free fall is a Minkowski space according to special relativity. What Einstein added to these theorems is that physically such a locally constant metric must be Minkowski. This is also the theoretical basis of the Einstein-Minkowski condition that Einstein uses to derive the bending of light rays and the gravitational redshifts [4, 5].

Thus, Pauli's version [24] is a simplified but corrupted version of these theorems as follows:

"For every infinitely small world region (i.e. a world region which is so small that the space- and time-variation of gravity can be neglected in it) there always exists a coordinate system K_0 (X_1, X_2, X_3, X_4) in which gravitation has no influence either

in the motion of particles or any physical process."

Pauli regards the equivalence principle as merely the existence of locally constant spaces. Then, Pauli's version is only a corrupted mathematical statement which may not be physically realizable because of the theorems.

A crucial error is that Pauli extended the removal of uniform gravity to the removal of gravity in a small region. This is simply incorrect in mathematics. Because many physicists do not understand mathematical analysis, they did not recognize that the removal of gravity in a small region, no matter how small, would be very different from a removal of gravity at one point. Apparently, neither Pauli [24] nor the Wheeler School [1, 2, 28, 29] understands the mathematics of the above theorems [27].

Misner et al. [1] claimed that Einstein's equivalence principle is as follows: "In any and every local Lorentz frame, anywhere and anytime in the universe, all the (nongravitational) laws of physics must take on their familiar special-relativistic form. Equivalently, there is no way, by experiments confined to infinitesimally small regions of spacetime, to distinguish one local Lorentz frame in one region of spacetime frame from any other local Lorentz frame in the same or any other region." They even claimed this as the Einstein's principle in its strongest form[7]. However, this version makes essentially another form of the misinterpretation of Pauli [24]. They do not seem to understand or to be aware of the related mathematics [27], and their followers probably have similar problems. *This version of the Wheeler School combines errors of Pauli and the 1911 assumption, but ignores the Einstein-Minkowski condition that is the physical essence of Einstein's principle.* [8]

Although Einstein's equivalence principle was clearly illustrated only recently [30-32], the Wheeler School should bear the responsibility of their misinformation on this principle [1] by ignoring both crucial work of Einstein, i.e., references [4] and [5], and related theorems [27], and giving an invalid version of such a principle. *A main problem is that the Einstein-Minkowski condition [4, 5], which plays a crucial role in measurement, is eliminated.*

3. Invalidity of Einstein's Covariance Principle and Einstein's Theory of Measurements

As shown by Zhou [33, 34], Einstein's equivalence principle is actually inconsistent with his covariance principle. Einstein's covariance principle is a source of errors that sustains misinterpretations [21, 33-35]. Starting from this "principle", Einstein implicitly assigns different phys- ical meaning to coordinates for different gauges [6, 36, 37].

The principle of general relativity states "The law of physics must be of such a nature that they apply to systems of reference in any kind of motion. Einstein extended this principle to unrestricted covariance and called it as the "principle of covariance" [4, 5]. He stated, "The general laws of nature are to be expressed by equations which hold good for all systems of co-ordinates, that is, are co-variant with respect to any substitutions whatever (generally co-variant)."

However, as Einstein [32] pointed out, the time coordinate must be distinct from a space coordinate. Moreover, the gauge conditions are known to be not tensor conditions. Einstein failed to see that different gauges would lead to different physical interpretations of the coordinates, but Zhou did [33, 34]. Based on that both the Schwarzschild and the harmonic solution produced the same first order deflection of a light ray, Einstein [5] prematurely remarked, "It should be noted that this result, also, of the theory is not influenced by our arbitrary choice of a system of coordinates."

In Einstein's arguments for this principle, he emphasized that a physical theory is about the coincidences of the space-time points, but the meaning of measurements is crucially omitted [30]. Eddington [38] commented, "space is not a lot of points close together; it is a lot of distances interlocked." To describe events, one must be able to relate events of different locations in a definite manner [39]. Moreover, as pointed out by Morrison of MIT[9], the "covariance principle" is invalid because it disrupts the necessary physical continuity from special relativity to general relativity [30, 39].

Note that Einstein's "principle of covariance" has no theoretical basis or observational support beyond that allowed by the principle of general relativity [39]. To start with, the covariance principle was proposed as a remedy for the deficiency of Einstein's adaptation of the notion of distance in a Riemannian space. Such an adaptation has been pointed out by Whitehead [40] as invalid in physics. Moreover, it is found that his

justifications for his adaption are due to invalid applications of special relativity [21].

Moreover, his calculation for the bending of light has actually proved that his theory of measurement is experimentally invalid. If one defines the distance as in the Riemannian space, one would get only half of the observed value of light blending [41]. It turns out, however, that the correct theory is just what Einstein practiced in his calculation of the bending of light [37]. Thus, the interpretation of Hubble law, based on Einstein's theory of measurement, as due to the Doppler effect is invalid [41] (see Appendix).

Nevertheless, many still believe in this invalid "principle", in part, because gauge invariance has a long history starting from electrodynamics. The notion of gauge invariance has been developed according to non-Abelian gauge theories such as the Yang-Mills-Shaw theory [42, 43] [10]. They naively extended the invariance of the Abelian gauge to the cases of the Non-Abelian gauges in terms of mathematics. However, as shown by Aharonov & Bohm [44], the electromagnetic potentials actually are physically effective; and, as shown by Weinberg [45], all the physical non-Abelian gauge theories are not gauge invariant such that masses can be generated. These facts support the view that gauge invariance of the whole theory would be a manifestation that there are some deficiencies [46, 47].

It has been shown by Bodenner & Will [48] and Gérard & Piereaux [49] that the deflection angle is gauge invariant to the second order. However, upon examining the physical meaning of the impact parameter b of the light ray and the shortest distance r_0 from the light ray to the center of the sun, it is clear that these physical quantities cannot be both gauge invariant. From the Schwarzschild gauge and the harmonic gauge, one has respectively

$$b \approx \kappa M + r_0, \qquad (4a)$$

but

$$b \approx 2\kappa M + r_0 . \qquad (4b)$$

Thus, Einstein's covariance principle is invalid in physics since two physical quantities should not have gauge dependent relations.

Another counter example for the covariance principle is the formulas for the de Sitter precession. From the Maxwell-Newton Approximation [7-9], one would obtain a formula [39] that is different from the formula obtained from the Kerr metric. However, they cannot be distinguished by the Stanford experiment, gravity Probe-B because this experiment detects only the time average. The time average of the difference is essentially zero. It seems a feasible simple experiment to show that the break down of gauge invariance is still the experiment on local light speeds [37] pioneered by Zhou [34].

Nevertheless, Misner et al. [1, p. 430] claimed that the covariance principle can be verified experimentally, but provided the opposite evidence. [11] For instance, Will [1; p. 1067] claimed Whitehead's theory is invalid; but the solution of Whitehead is diffeomorphic to Einstein's [50]. Their motivation seems to justify such an invalid "principle" because it is often used in arguments of their theory of black holes. One may wonder why nobody corrected their mistake. The answer would be that that many theorists often failed to distinguish the difference between physics and mathematics [3].

Moreover, since the covariance principle is necessary to remedy the shortcomings of Einstein's theory of measurement [5], which was justified with applications of special relativity, many would still believe in the covariance principle even though counter examples have been found [35]. *Thus, to understand the issue of the covariance principle thoroughly, one must also examine Einstein's justification for "measurement" with applications of special relativity* [21].

In the book of Misner et al., their errors in physics, mathematics and logic are exposed, but were not recognized. This supports the claim of Feynman [51] that many theorists in gravitation are just incompetent. The root of such errors is due to a failure to distinguish the difference between mathematics and physics. To see all these errors clearly, it is necessary to understand also the principle of causality for which physicists would understand.

4. The Principle of Causality and the Einstein Equation

The time-tested assumption that phenomena can be explained in terms of identifiable causes is called the principle of causality [7-9]. This principle is the basis of relevance for all scientific investigations, and thus is always implicitly used [50]. This principle is commonly used in symmetry considerations in electrodynamics.

Einstein and other theorists have used this principle implicitly on symmetry considerations [8] such as for a circle in a uniformly rotating disk and the metric for a spherically symmetric mass distribution. Nevertheless,

this principle is often neglected [9, 52] because the confusion on physical coordinates created by the invalid covariance principle that would make it almost impossible to justify the symmetry used. *Applications of the principle of causality become clear after Einstein's equivalence principle is understood* [30, 53].

Because of the "covariance principle", the coordinates were ambiguous, and thus it is often difficult to apply the principle of causality in a logical manner other than implicitly as Einstein did. Since the covariance principle is necessary to remedy the shortcomings of Einstein's theory of measurement [5], many would give up only after it was found recently that the justifications of Einstein's theory of measurement actually were based on invalid applications of special relativity [21, 54], in addition to being in disagreement with observed bending of light rays.

There are other useful consequences. For instance, the weak sources would produce weak gravity is the foundation of Einstein's requirement on weak gravity [8]. The unbounded "weak waves" of Bondi, Pirani, & Robinson [55] are not valid because it cannot be reduced to the flat metric when gravity is absent. Parameters unrelated to any physical cause in a solution are not allowed. For instance, Penrose [56] accepted the metric with an electromagnetic plane-wave as a source, but it is not valid in physics because unphysical parameters are involved [57]. Moreover, a dynamic solution must be related to appropriate dynamic sources [58].

One might argue that a gravitational plane-wave would have no source. For the fact that a plane-wave is intrinsically unbounded, there is no valid explanation until the principle of causality is recognized. A plane wave is not real - but a local idealization of a section of the wave. For a cylindrical symmetric wave, however, appropriate sources must be present. The Einstein-Rosen type waves are invalid because it is impossible to have physically appropriate sources [58]. However, the principle of causality can be misunderstood.

For instance, 't Hooft naively claimed [59], "Dynamical solutions means solutions that depend non-trivially on space as well as time. Numerous of such solutions are being generated routinely in research papers ..." Thus, he has different, but invalid understanding of the principle of causality. He [59] claimed, "To me, causality means that the form of the data in the future, $t > t_1$, is completely and unambiguously dictated by their values and, if necessary, time derivatives in the past, $t = t_1$. So, I constructed the complete Green function for this system and showed it to Mr. L. This function gives the solution at all times, once the solution and its first time derivative is given at $t = t_1$, which is a Cauchy surface." However, his data actually are calculated values only [59] [12).

Thus, his causality only means that a Maxwell-type equation, which produces the Green function, is satisfied. This is inadequate because a solution of the Maxwell equation could violate the principle of causality. For instance, the electromagnetic potential $A_0[\exp(t - z)^2]$ (A_0 is a constant), is invalid in physics. Although a plane-wave can be considered as an idealization of a field generated by sources, this function cannot be considered as such an idealization [58].

In defense of the errors of the Nobel Committee for Physics of 1993, 't Hooft [60] comes up with a bounded time-dependent cylindrical symmetric solution as follows:

$$\Psi(r,t) = A \int_0^{2\pi} d\varphi\, e^{-\alpha(t-r\cos\varphi)^2}, \qquad (5)$$

where A and $\alpha > 0$ are free parameters. $|\Psi|$ is everywhere bounded. However, it has been shown that there are no valid sources that can be related to solution (5) [60]. Thus, since the principle of causality is also violated, his solution is not valid in physics.

It should be noted that while 't Hooft is a very good applied mathematician, his understanding of physics is far from adequate. From the above example, it is clear that 't Hoof does not understand electrodynamics and special relativity adequately. Moreover, from his Nobel Lecture [61] it also clear that he does not understand special relativity and even Newtonian mechanics adequately. 't Hooft claimed in his Nobel Lecture that the electric energy is part of the physical mass m_{phys} of an electron. Moreover, he claimed the "physical mass" obeys Newton's second law $F = m_{phy}\, a$. Note that such a claim violates special relativity because part of the electric energy is far from the electron, and thus cannot react immediately as an inertial mass does.

Many relativists recognize the light speed as the speed limit of physical influence, but failed to understand the principle of causality. Moreover, the covariance principle would confuse applied mathematicians such as 't Hooft, [12) to fail in distinguishing physics from mathematics [59]. In fact, journals such as the Physical Review also do not understand the principle of causality adequately, and accept unbounded solutions [58]. However, since a bounded dynamic solution is needed for the calculation of gravitational radiation, *the non-existence of a bounded dynamic solution remains an unsolved issue.*

Based on his field equation, Einstein [4, 5] made three predictions namely: 1) the gravitational redshifts, 2) the perihelion of Mercury, and 3) the deflection of light. Observations accurately confirm and create a faith in his theory. However, these confirmations are actually inflated and explained as follows:

1) The gravitational redshifts were first derived from the invalid 1911 assumption of the equivalence between acceleration and Newtonian gravity. This shows that the gravitational redshifts can be derived from an invalid theory.

2) The observed bending of light is inconsistent with Einstein's theory of measurement [60], [13) but is consistent with the measurement based on the Euclidean-like structure if his equivalence principle is valid for the metric [5].

3) As Gullstrand [62] suspected, in 1995 it has been proven impossible to have a bounded dynamic solution. [14) Thus, the perihelion of Mercury, in principle, is still beyond the reach of the Einstein equation [7]. *This fundamental mistake in calculation, as will be shown, has far reaching influences to other important errors in astrophysics.*

Also, Einstein's controversial notion of gravitational energy-stress being a pseudo-tensor has been proven incorrect [7]. Since Einstein's covariance principle is proven to be invalid [6], and diffeomorphic solutions with the same frame of reference are not equivalent in physics. Therefore, actually none of the predictions had a solid theoretical foundation yet.

An urgent issue is to find a valid physical gauge for a given problem. Fortunately, the Maxwell-Newton approximation has been proven to be an independently valid first order approximation for gravity due to massive sources [8], so that the binary pulsar radiation experiments can be explained satisfactorily [7-9]. Thus, Einstein's notion of weak gravity (including gravitomagnetism and gravitational radiation [63]) is valid [8, 57] [15). Moreover, calculations of the Hulse-Taylor experiments of the binary pulsars necessitate that the coupling constants have different signs [7]. Thus, the assumption of a unique coupling sign for the singularity theorems [2] of Penrose and Hawking is proven invalid. [16)]

5. The Nonexistence of Dynamic Solutions for the Einstein Equation

In general relativity, the most important issue is whether dynamic solutions exist for the Einstein equation. The issue existed from the beginning of this theory until currently. The question started with the calculation of the perihelion of Mercury. In 1915 Einstein obtained the expected value of the remaining perihelion with his equation, and thus was confident of its correctness. The subsequent confirmation of the bending of light further boosted his confidence. However, unexpectedly the base of his confidence was questioned by Gullstrand [62], the Chairman (1922-1929) of the Nobel Prize for Physics. The perihelion of Mercury is actually a many-body problem, but Einstein had not shown that his calculation could be derived from such a necessary step. Thus, Mathematician D. Hilbert, who approved Einstein's initial calculation, did not come to its defense.

In spite of objections from many physicists, Gullstrand stayed on his position, and Einstein was awarded a Nobel Prize by virtue of his photoelectric effects instead of general relativity as expected. The fact is, however, Gullstrand is right. In 1995, it is proven that Einstein's equation is incompatible with gravitational radiation and also does not have a dynamic solution [7-9].

For space-time metric $g_{\mu\nu}$, the Einstein equation of 1915 [5] is

$$G_{\mu\nu} \equiv R_{\mu\nu} - \frac{1}{2} g_{\mu\nu} R = -KT(m)_{\mu\nu} \qquad (6)$$

where $G_{\mu\nu}$ is the Einstein tensor, $R_{\mu\nu}$ is the Ricci curvature tensor, $T(m)_{\mu\nu}$ is the energy-stress tensor for massive matter, and K ($= 8\pi\kappa c^{-2}$, and κ is the Newtonian coupling constant) is the coupling constant.· Thus,

$$G_{\mu\nu} \equiv R_{\mu\nu} - \frac{1}{2} g_{\mu\nu} R = 0, \quad \text{or} \qquad R_{\mu\nu} = 0, \qquad (6')$$

at vacuum. However, (6') also implies no gravitational wave to carry away energy-momentum in vacuum [64] and thus the principle of causality is violated [65]. Currently a popular error is to claim that general relativity has totally superseded Newtonian gravity [66]. However, this is not yet true because general relativity does not have a solution for a two-body problem [67].

There are serious consequences for the error of the mistaken existence of dynamic solutions for the 1915 Einstein equation. Since the equation is incorrect for the dynamic case; it would lead to erroneous conclusions [3]. A well-known result is the existence of the so-called space-time singularities due to Penrose and Hawking by implicitly assuming the unique sign for all the coupling constants [2]. Another result is that the correct conjecture of Einstein on unification between gravitation and electromagnetism was not recognized [10]. Thus, the criticism of Gullstrand turns out to be very crucially constructive and beneficial.

Nevertheless, following Einstein's error, there are many erroneous claims for the existence of a dynamic solution because of mathematical inadequacy among physicists. Some mathematicians such Yau [68] and Witten [69] implicitly used the existence of bounded dynamic solution but were unaware of its invalidity [70] because of inadequate background in physics. Moreover, such claims are not only accepted by the 1993 Nobel Prize Committee for Physics but also D. Christodoulou was awarded the 2011 Shaw Prize[17)] in mathematics to honor for his errors against the honorable Gullstrand [62] [18)]. Subsequently, Christodoulou was accepted as a member of U.S. National Academy of Sciences (2012). Due to inadequacy in mathematics and physics, theorists make serious errors in addition to Einstein's limitation. Thus, generations of physicists are misled into serious errors because of unprecedented over confidence [6].

In addition to Einstein's error, there are popular mathematical errors among physicists that lead to the present sad situation. Many believed that the non-linear Einstein equation must have a bounded dynamic solution since the linearized Einstein equation has bounded dynamic solutions. [19)] Also, this has been proven true for the static case. Unfortunately, Einstein and his peers did not see that for a non-linear equation, these equations can actually be independent equations for the dynamic case [71]. It is also unfortunate that mathematicians also help in perpetuating the errors because they do not understand the physics of a nonlinear equation [6].

To help the readers, we shall first give some examples that illustrate the non-existence of dynamic solutions. Moreover, the crucial errors and blind spots of current theorists are pointed out. Note that the non-existence of bounded plane-wave solutions as illustrated by Bondi et al. [55] is evidence that there is no bounded wave solution. It is hoped that this will help theorists to pay attention to the proof [7-9] for the crucial problem of the non-existence of dynamic solutions.

5.1. *Examples of no Bounded Dynamic Solutions*

Here, we shall show with examples to illustrate that for the case of gravitational waves, there is no bounded dynamic solution. Now, consider another well-known metric obtained by Bondi, Pirani, & Robinson [55] as follows:

$$ds^2 = e^{2\varphi}\left(d\tau^2 - d\xi^2\right) - u^2\left[\begin{array}{l}\cosh 2\beta\left(d\eta^2 + d\varsigma^2\right)\\ +\sinh 2\beta \cos 2\theta\left(d\eta^2 - d\varsigma^2\right)\\ -2\sinh 2\beta \sin 2\theta\, d\eta\, d\varsigma\end{array}\right] \quad (7a)$$

where ϕ, β and θ are functions of $u\ (=\tau-\xi)$. It satisfies the differential equation (i.e., their Eq. [2.8]),

$$2\phi' = u\left(\beta'^2 + \theta'^2 \sinh^2 2\beta\right) \quad (7b)$$

They claimed this is a wave from a distant source. (7b) implies ϕ *cannot be a periodic function.* The metric is irreducibly unbounded because of the factor u^2. Eq. (7b) is a special cases of $G_{\mu\nu} = 0$. However, linearization of (7b) does not make sense since variable u is not bounded. They claimed the notion of weak gravity invalid because they do not understand the principle of causality adequately.

Moreover, when gravity is absent, it is necessary to have $\phi = \sinh 2\beta = \sin 2\theta = 0$. These would reduce (7a) to

$$ds^2 = \left(d\tau^2 - d\xi^2\right) - u^2\left(d\eta^2 - d\varsigma^2\right) \quad (7c)$$

However, this metric is not equivalent to the flat metric. Thus, metric (7c) violates the principle of causality. Also it is impossible to adjust metric (7a) to become equivalent to the flat metric.

This challenges the view that both Einstein's notion of weak gravity and his covariance principle are valid. These conflicting views are supported respectively by the editors of the "Royal Society Proceedings A" and the "Physical Review D"; thus there is no general consensus. As the Royal Society correctly pointed out [55, 72], Einstein's notion of weak gravity is inconsistent with his covariance principle. Note that Einstein's covariance principle has been proven invalid since counter examples have been found [6, 39].

There are other theorists who also ignore the principle of causality. Consider another "plane wave", which is intrinsically non-physical, is the metric accepted by Penrose [56] as follows:

$$ds^2 = du\, dv + Hdu^2 - dx_i dx_i \text{ ,where } H = h_{ij}(u)x_i x_j \quad (8)$$

where $u = ct - z$, $v = ct + z$. However, there are non-physical parameters (the choice of origin) that are unrelated to any physical causes. Being a mathematician, Penrose [56] over-looked the principle of causality. Also, linearization of metric (8) does not make sense since its metric elements are unbounded.

Another good example is the plane-wave solution of Liu & Zhou [73], which satisfies the harmonic gauge, is as follows:

$$ds^2 = dt^2 - dx^2 + 2F(dt - dx)^2 - \cosh 2\psi(e^{2\phi}\,dy^2 + e^{-2\phi}dz^2) - 2\sinh 2\psi\, dy\, dz. \quad (9)$$

where $\phi = \phi(u)$ and $\psi = \psi(u)$. Moreover, $F = F_P + H$, where

$$F_P = \frac{1}{2}(\dot{\psi}^2 + \dot{\phi}^2 \cosh^2 2\psi)\,[\cosh 2\psi\,(e^{2\phi}\,y^2 + e^{-2\phi}\,z^2)$$
$$+2\sinh 2\phi\,yz], \quad (10)$$

and H satisfies the equation,

$$\cosh 2\psi\,(e^{-2\phi}H_{,22} + e^{2\phi}\,H_{,33}) - 2\sinh 2\psi\,H_{,23} = 0. \quad (11)$$

For the weak fields one has $1 \gg |\phi|$, $1 \gg |\psi|$, but there is no weak approximation as claimed to be

$$ds^2 = dt^2 - dx^2 - (1 + 2\phi)\,dy^2 - (1 - 2\phi)\,dz^2 - 4\psi dy\, dz \quad (12)$$

because F_p is not bounded unless $\dot{\phi}$ and $\dot{\psi}$ are zero (i.e., no wave).

The linearized equation for a dynamic case has been illustrated as incompatible with the non-linear Einstein equation, which has no bounded dynamic solutions. Thus, Eq. (7b), Eq, (8), and Eq. (12) serve as good simple examples that can be shown through explicit calculation that linearization of the Einstein equation is not valid. Also, metric (8) suggests that the cause of having no physical solution would be due to inadequate source terms [7, 74].

An independent supplementary convincing evidence for the absence of a bounded dynamic solution is, as shown by Hu, Zhang & Ting [75], that a calculated gravitational radiation would depend on the approach used. This manifests that there is no bounded solution [76]. Also, approximation schemes such as post-Newtonian approximation [77] are that their validity is also assumed only.

5.2. *Errors of Wald*

In the pretext of a "modern view", R. M. Wald [2] rejected Einstein's equivalence principle. Wald [2] claimed the equivalence of inert mass and the gravitational mass due to Galileo and Newton as "the equivalence principle". However, the 1993 Nobel Prize Committee for Physics adopted this view because of inadequate understanding of the equivalence principle that Einstein emphasizes in his life time. In so doing, Wald also avoided criticizing Misner et al. [1]. Clearly Wald does not understand Einstein's equivalence principle, which plays a crucial role in establishing the validity of the Maxwell-Newton Approximation independently [8]. Apparently, he chose Einstein's covariance principle because he did not see that it is invalid in physics.

Wald [2], has also been mistaken on dynamic solutions. In general relativity, weak sources would produce a weak field, i.e.,

$$g_{\mu\nu} = \eta_{\mu\nu} + \gamma_{\mu\nu}, \text{ where } |\gamma_{\mu\nu}| \ll 1 \qquad (13)$$

and $\eta_{\mu\nu}$ is the flat metric. However, according to the principle of causality, (13) should be valid; but only if the equation is valid in physics. In other words, condition (13) for weak gravity and whether the principle of causality is applicable need a rigorous proof.

Unfortunately, many believe that condition (13) for weak gravity is always valid because of accurate predictions for the static case. When the Einstein equation has a weak solution, an approximate weak solution can be derived through the approach of the field equation being linearized. The linearized Einstein equation with the linearized harmonic gauge $\partial^\mu \bar{\gamma}_{\mu\nu} = 0$ is

$$\frac{1}{2}\partial^\alpha \partial_\alpha \bar{\gamma}_{\mu\nu} = \kappa T_{\mu\nu} \text{ where } \bar{\gamma}_{\mu\nu} = \gamma_{\mu\nu} - \frac{1}{2}\eta_{\mu\nu}\gamma$$

and $\gamma = \eta^{\alpha\beta}\gamma_{\alpha\beta}$. $\qquad (14)$

Note that we have

$$G_{\mu\nu}^{(1)} = \frac{1}{2}\partial^\alpha \partial_\alpha \bar{\gamma}_{\mu\nu} - \partial^\alpha \partial_\mu \bar{\gamma}_{\nu\alpha} - \partial^\alpha \partial_\nu \bar{\gamma}_{\mu\alpha} + \frac{1}{2}\eta_{\mu\nu}\partial^\alpha \partial^\beta \bar{\gamma}_{\alpha\beta}$$

and $\quad G_{\mu\nu} = G_{\mu\nu}^{(1)} + G_{\mu\nu}^{(2)}$. $\qquad (15)$

The linearized vacuum Einstein equation means

$$G_{\mu\nu}^{(1)}[\gamma_{\alpha\beta}^{(1)}] = 0 \qquad (16)$$

Thus, as pointed out by Wald, in order to maintain a solution of the vacuum Einstein equation to second order we must correct $\gamma^{(1)}_{\mu\nu}$ by adding to it the term $\gamma^{(2)}_{\mu\nu}$, where $\gamma^{(2)}_{\mu\nu}$ satisfies

$$G_{\mu\nu}^{(1)}[\gamma^{(2)}{}_{\alpha\beta}] + G_{\mu\nu}^{(2)}[\gamma_{\alpha\beta}] = 0, \text{ where } \gamma_{\mu\nu} = \gamma^{(1)}{}_{\mu\nu} + \gamma^{(2)}{}_{\mu\nu} \quad (17)$$

which is the correct form of eq. (4.4.52) in [2] (Wald did not distinguish $\gamma_{\mu\nu}$ from $\gamma^{(1)}{}_{\mu\nu}$). However, detailed calculation shows that this equation does not have a solution for the dynamic case [7-9] although it does have a solution for the static case.

If there is no solution for eq. (9), then the Einstein equation does not have a bounded dynamic solution [76]. In conclusion, due to confusion between mathematics and physics, Wald [2] also made errors in mathematics at the undergraduate level. The principle of causality requires the existence of a dynamic solution, but Wald did not see that the Einstein equation can fail this requirement.

5.3. *Errors of Misner, Thorne and Wheeler*

Currently the Wheeler School led by Misner, Thorne and Wheeler [1] is probably the most influential. Unfortunately, this school not only misinterpreted and distorted general relativity, but also makes the error of claiming the existence of dynamic solutions.

A "wave" form considered by Misner, Thorne, & Wheeler [1] is as follows:

$$ds^2 = c^2 dt^2 - dx^2 - L^2\left(e^{2\beta}dy^2 + e^{-2\beta}dz^2\right) \qquad (18)$$

where $L = L(u)$, $\beta = \beta(u)$, $u = ct - x$, and c is the light speed. Then, the Einstein equation $G_{\mu\nu} = 0$ becomes

$$\frac{d^2 L}{du^2} + L\left(\frac{d\beta}{du}\right)^2 = 0 \qquad (19)$$

Misner et al. [1] claimed that Eq. (19) has a bounded solution, compatible with a linearization of metric (18).

It has been shown with undergraduate mathematics [53] that Misner et al. are incorrect and Eq. (19) does not have a physical solution that satisfies Einstein's requirement on weak gravity. In fact, $L(u)$ is unbounded even for a very small $\beta(u)$.

On the other hand, from the Maxwell-Newton approximation in vacuum, Einstein [78] obtained a solution as follows:

$$ds^2 = c^2 dt^2 - dx^2 - (1+2\phi)dy^2 - (1-2\phi)dz^2 \qquad (20)$$

where ϕ is a bounded function of u $(= ct - x)$. Note that metric (20) is the linearization of metric (18) if $\phi = \beta(u)$. Thus, the problem of waves illustrates that the linearization may not be valid for the dynamic case when gravitational waves are involved since eq. (19) does not have a weak wave solution. Since this crucial calculation can be proven with mathematics at the undergraduate level, it should not be surprising that Misner et al. [1] make other serious errors in mathematic and physics such as on the local time in their eq. (40.14).

The root of the errors of Misner et al. was that they incorrectly [7-9] assumed that a linearization of a non-linear equation would always produce a valid approximation. Linearization of (19) yields $L'' = 0$, and in turn this leads to $\beta'(u) = 0$. In turn, this leads to a solution $L = C_1 u + 1$ where C_1 is a constant. Therefore, if $C_1 \neq 0$, it contradicts the requirement $L \approx 1$. Moreover, $\beta'(u) = 0$. implies no wave. Thus, one cannot get a weak wave solution through linearization of Eq. (19), which has no bounded solution. This shows also that the assumption of metric form (18) [1], which has a weak form (20), is not valid for the Einstein equation.

Many regard a violation of the Lorentz symmetry also as a violation of general relativity. However, this notion actually comes from the distortion of Einstein's equivalence principle by Misner et al. [1] (see Section 2). Apparently, they do not understand or probably even were unaware of the related mathematics [27]. Moreover, in their eq. (40.14) they got an incorrect local time of the earth in disagreement with Wald [2] and Weinberg [77]. Clearly they [1] failed to understand Einstein's equivalence principle [4, 5].

Furthermore, Thorne [79] criticized Einstein's principle with his own distortion as follows:

"In deducing his principle of equivalence, Einstein ignored tidal gravitation forces; he pretended they do not exist. Einstein justified ignoring tidal forces by imagining that you (and your reference frame) are very small."

Perhaps, Thorne did not know the fact that *the term "Einstein elevator" is due to Bergmann but not Einstein.*

Although Einstein's equivalence principle was clearly illustrated only recently, the Wheeler School should bear the responsibility for misinformation on this principle (see Section 2). Moreover, they give *the irrelevant incorrect 1911 assumption on the equivalence of Newtonian gravity and acceleration as references* although Newtonian gravity is proven not equivalent to acceleration.

5.4. *The Errors of Christodoulou*

The fact that Christodoulou received honors for his errors related to the Einstein equation testified,

"Unthinking respect for authority is the greatest enemy of truth" as Einstein asserted. The strategy of the Nobel Prize based on the recognition time lag failed because mathematical and logical errors can be subtle. Many theorists just do not have the caution, and patience and/or the mathematical background to find out the subtle errors involved as shown in the press release of the Nobel Committee [15].

Due to errors in undergraduate mathematics [80], Christodoulou & Klainerman [81] claimed that they have constructed dynamic solutions. However, one should not be too surprised because Christodoulou obtained his Ph. D. under Wheeler, who also has similar problems in mathematics (see Section 4.3). Because of the support of the Princeton University, progress in physics did suffer not only from their errors, but also wasting the resources. Fortunately such a struggle comes to an end when their errors can be illustrated with mathematics at the undergraduate level [80]. Moreover, only after the non-existence of a dynamic solution for the Einstein equation was recognized, Einstein's conjecture of the unification can be proven correct [10, 82, 83].

The book of Christodoulou & Klanerman [81] is confusing (see also [82]). Their main Theorem 1.0.3 states that any strongly asymptotically flat (S.A.F.) initial data set that satisfies the global smallness assumption leads to a unique globally hyperbolic asymptotically flat development. However, because the global smallness assumption has no dynamic requirements in their proofs, there is no assurance for the existence of a dynamic S.A.F. initial data set [80]. Thus, the existence of a bounded dynamic initial set is assumed only, and their proof is at least incomplete.

Perlik [84] commented, "What makes the proof involved and difficult to follow is that the authors introduce many special mathematical constructions, involving long calculations, without giving a clear idea of how these building-blocks will go together to eventually prove the theorem. The introduction, almost 30 pages long, is of little help in this respect. Whereas giving a good idea of the problems to be faced and of the basic tools necessary to overcome each problem, the introduction sheds no light on the line of thought along which the proof will proceed for mathematical details without seeing the thread of the story. This is exactly what happened to the reviewer." Essentially, they assume the existence of a bounded initial set to prove the existence of a bounded solution. *Moreover, his initial condition has not been proven as compatible with the Maxwell-Newton approximation which is valid for weak gravity* [80].

This book review originally appeared in ZfM [84] in 1996; and republished in the journal, GRG [85] again with the editorial note, "One may extract two messages: On the one hand, (by seeing e.g. how often this book has been cited), the result is in fact interesting even today, and on the other hand: There exists, up to now no generally understandable proof of it." The review actually suggests that problems would be adequately identified in the introduction.

As shown in reference [80], the possible nonexistence of their dynamic solutions and its incompatibility with Einstein's radiation formula can be discovered in their introduction. From this review, what the Shaw Prize claimed as "for their highly innovative works on nonlinear partial differential equations in Lorentzian and Riemannian geometry and their applications to general relativity and topology.", in the case of Christodoulou, seems to be just a euphemism for a highly confusing and incomprehensible presentation. These, especially the confusion in the introduction, manifest that the authors have not grasped the essence of the problem.

Moreover, they seem to try to create enough confusion, as Misner et al. [1] did on their plane-wave solution for eq. (19), to gain the acceptance from the readers, with the support of the Princeton University. However, careful calculation with undergraduate mathematics shows that, just as Misner et al. [1] are wrong [53], Christodoulou is also incorrect [80].

Many theorists assume a physical requirement would be unconditionally satisfied by the Einstein equation [86], and such a view was adapted by Christodoulou. According to the principle of causality, a bounded dynamic solution should exist, but this does not necessarily mean that the Einstein equation has such a solution. As shown, the mathematical analysis of Christodoulou is also not reliable at the undergraduate level although he claimed to have such a strong interest in his autobiography.

Gullstrand was not the only theorist who questioned the existence of the bounded dynamic solution for the Einstein equation. As shown by Fock [87], any attempt to extend the Maxwell-Newton approximation to higher approximations leads to divergent terms. In 1995, it has

been proven [7-9] that for a dynamic case the linearized equation as a first order approximation, is incompatible with the nonlinear Einstein field equation. Moreover, the Einstein equation does not have a dynamic solution for weak gravity unless the gravitational energy tensor with an anti-gravity coupling is added to the source. The necessity of an anti-gravity coupling term manifests why a bounded wave solution is impossible for Einstein's equation (see Next Section).

Their book [81] was accepted because it supports and is consistent with existing errors as follows:

1) It supports errors that created a faith in the existence of dynamical solutions of physicists including Einstein etc.
2) Due to the inadequacy of the mathematics used, the book was cited before 1996 without referring to the details.
3) Nobody suspected that professors in mathematics and/or physics could make mistakes at the <u>undergraduate</u> level.
4) Because physical requirements were not understood, unphysical solutions were accepted as valid [6, 87-89].

Thus, in the field of general relativity, strangely there is no expert almost 100 years after its creation.

In physics, a dynamic solution must be related to dynamic sources, but a "time-dependent" solution may not necessarily be a physical solution [55, 90, 91]. For instance, their "initial data sets" can be incompatible with the field equation for weak gravity. Second, the only known cases are static solutions. Third, they have not been able to relate any of their constructed solutions to a dynamic source. In pure mathematics, if no example can be given, such abstract mathematics is likely wrong [92].

In fact, in 1953 Hogarth [93] conjectured that a dynamic solution for the Einstein equation does not exist. Moreover, in 1995 it is proven impossible to have a bounded dynamic solution because the principle of causality is violated [7-9].

Nevertheless, the mistakes of the 1993 Nobel Committee probably show that the level of misunderstanding in general relativity then that had led to a number of awards and honors for the errors of D. Christodoulou (Wikipedia) as follows:

MacArthur Fellows Award (1993);
Bôcher Memorial Prize (1999);
Member of American Academy of Arts and Sciences (2001);

Tomalla Foundation Prize (2008);
Shaw Prize (2011);
Member of U.S. National Academy of Sciences
(2012).

Note that there are many explicit examples that show the claims of Christodoulou are incorrect [55, 75, 82]. However, due to the practice of biased authority worship, many theorists just ignored them. Physically, a bounded dynamic solution should exist, but Einstein's field equation just does not have such a solution. Now, in view of the facts that Christodoulou's contributions to general relativity are essentially just errors, it is up to the U.S. National Academy of Sciences to handle such a special case.

Note that their book [81] has been criticized by Volker Perlick [84, 85] as "incomprehensible". Moreover, S. T. Yau has politely lost his earlier interests on their claims [81]. The awards and honors to Christodoulou manifested an unpleasant fact that most of the physicists do not understand pure mathematics adequately and many mathematicians do not understand physics. It should be noted also that J. A. Taylor failed to justify his calculation on the binary pulsars when P. Morrison of MIT questioned him [64].

6. The Gravitational Wave and the Anti-gravity Coupling in the Einstein Equation

From the above discussions, it is clear that a dynamic solution of the Einstein equation would not be bounded. Here, it will be shown that this is always true when a gravitational wave would be present. This is obviously in conflict with that the linearized Einstein equation has bounded dynamic solutions. It turns out that, for the dynamic case, the linearized Einstein equation is a linearization of a modified Einstein equation, but is independent of the Einstein equation [71].

First, a major problem is a mathematical error on the relationship between Einstein equation and its "linearization". It was incorrectly believed that the linear Maxwell-Newton Approximation [7] (or the linearized Einstein equation [5])

$$\frac{1}{2}\partial^c\partial_c \bar{\gamma}_{\mu\nu} = - K\, T(m)_{\mu\nu},$$

where $\bar{\gamma}_{\mu\nu} = \gamma_{\mu\nu} - \frac{1}{2}\eta_{\mu\nu}(\eta^{cd}\gamma_{cd})$ (21a)

and

$$\bar{\gamma}_{\mu\nu}(x^i, t) = - \frac{K}{2\pi}\int \frac{1}{R} T_{\mu\nu}[y^i, (t - R)]d^3y,$$

where $R^2 = \sum_{i=1}^{3} (x^i - y^i)^2$.   (21b)

provides the first-order approximation for the Einstein equation (6). This belief was verified for the static case only.

For a dynamic case, however, this is no longer valid. Note that the Cauchy data cannot be arbitrary for eq. (6). The Cauchy data of eq.(6) must satisfy four constraint equations, $G_{\mu t} = -KT(m)_{\mu t}$ ($\mu = x, y, z, t$) since $G_{\mu t}$ contains only first-order time derivatives [76]. This shows that (21a) would be dynamically incompatible with equation (6) [94]. [19] Further analysis shows that, in terms of both theory [8] and experiments [7], this mathematical incompatibility is in favor of (21), instead of (6).

In 1957, Fock [87] pointed out that, in harmonic coordinates, there are divergent logarithmic deviations from expected linearized behavior of the radiation. This was interpreted to mean merely that the contribution of the complicated nonlinear terms in the Einstein equation cannot be dealt with satisfactorily following this method and that another approach is needed. Subsequently, vacuum solutions that do not involve logarithmic deviation were founded by Bondi, Pirani & Robinson [55] in 1959. Thus, the incorrect interpretation appears to be justified and the faith on the dynamic solutions maintained. It was not recognized until 1995 [7] that such a symptom of divergence actually shows the absence of bounded physical dynamic solutions.

In physics, the amplitude of a wave is related to its energy density and its source. Equation (21) shows that a gravitational wave is bounded and is related to the dynamic of the source. These are useful to prove that (21), as the first-order approximation for a dynamic problem, is incompatible with equation (6). Its existing "wave" solutions are unbounded and therefore cannot be associated with a dynamic source [8]. In other words, there is no evidence for the existence of a physical dynamic solution.

With the Hulse-Taylor binary pulsar experiment [95], it became easier to identify that the problem is in (6). Subsequently, it has been shown that (21), as a first-order approximation, can be derived from physical requirements which lead to general relativity [8]. Thus, (21) is on solid theoretical ground and general relativity remains a viable theory. Note, however, that the proof of the nonexistence of bounded dynamic solutions for (6) is essentially independent of the experimental supports for (21).

To prove this, it is sufficient to consider weak gravity since a physical solution must be compatible with Einstein's [5] notion of weak gravity. To calculate the radiation, consider,

$$G_{\mu\nu} \equiv G^{(1)}{}_{\mu\nu} + G^{(2)}{}_{\mu\nu} \,,$$

where $\quad G^{(1)}{}_{\mu\nu} = \frac{1}{2}\partial^c\partial_c \bar{\gamma}_{\mu\nu} + H^{(1)}{}_{\mu\nu},$ (22a)

$$H^{(1)}{}_{\mu\nu} \equiv -\frac{1}{2}\partial^c[\partial_\nu \bar{\gamma}_{\mu c} + \partial_\mu \bar{\gamma}_{\nu c}] + \frac{1}{2}\eta_{\mu\nu}\partial^c\partial^d \bar{\gamma}_{cd} \,,$$

and $\quad 1 >> |\gamma_{\mu\nu}|.$ (22b)

$G^{(2)}{}_{\mu\nu}$ is at least of second order in terms of the metric elements. For an isolated system located near the origin of the space coordinate system, $G^{(2)}{}_{\mu t}$ at large r ($= [x^2 + y^2 + z^2]^{1/2}$) is of $O(K^2/r^2)$ [1, 2, 77].

One may obtain some general characteristics of a dynamic solution for an isolated system as follows:

1) The characteristics of some physical quantities of an isolated system:

For an isolated system consisting of particles with typical mass $\overline{M}$, separation $\bar{r}$, and velocities $\bar{v}$, Weinberg [77] estimated, the power radiated at a frequency ω of order $\bar{v}/\bar{r}$ will be of order

$$P \approx \kappa(\bar{v}/\bar{r})^6 \overline{M}^2 \bar{r}^4 \text{ or } P \approx \overline{M} \bar{v}^8/\bar{r} \,, \quad (23)$$

since $\kappa \overline{M}/\bar{r}$ is of order $\bar{v}^2$. The typical deceleration $\bar{a}_{rad}$ of particles in the system owing this energy loss is given by the power P divided by the momentum $\overline{M}\bar{v}$, or $\bar{a}_{rad} \approx \bar{v}^7/\bar{r}$. This may be compared with the accelerations computed in Newtonian mechanics, which are of order $\bar{v}^2/\bar{r}$, and with the post-Newtonian correction of $\bar{v}^4/\bar{r}$. Since radiation reaction is smaller than the post-Newtonian effects by a factor $\bar{v}^3$, if $\bar{v} << c$, the velocity of light, the neglect of radiation reaction is perfectly justified. This allows us to consider the motion of a particle in an isolated system as almost periodic.

Consider two particles of equal mass m with an almost circular orbit in the x-y plane whose origin is the center of the circle (i.e., the orbit of a particle is a circle if radiation is neglected). Thus, the principle of causality [74, 96] implies that the metric $g_{\mu\nu}$ is weak and very close to the flat metric at distance far from the source and that $g_{\mu\nu}$ (x, y, z, t') is an almost periodic function of t' (= t - r/c).

2) The expansion of a boun[19]ded dynamic solution g $_{\mu\nu}$ for an isolated weak gravitational source:

According (21), a first-order approximation of metric $g_{\mu\nu}$(x, y, z, t') is bounded and almost periodic since $T_{\mu\nu}$ is. Physically, the principle of causality requires $g_{\mu\nu}$ to be almost periodic in time since the motion of a source particle is. Such a metric $g_{\mu\nu}$ is asymptotically flat for a large distance r, and the expansion of a bounded dynamic solution is:

$$g_{\mu\nu}(n^x, n^y, n^z, r, t') =$$

$$\eta_{\mu\nu} + \sum_{k=1}^{\infty} f_{\mu\nu}{}^{(k)}(n^x, n^y, n^z, t')/r^k,$$

where $n^\nu = x^\nu/r$. (24a)

3) The non-existence of dynamic solutions:

It follows expansion (24a) that the non-zero time average of $G^{(1)}{}_{\mu t}$ would be of $O(1/r^3)$ due to

$$\partial_\mu n^\nu = (\delta^\nu{}_\mu + n^\nu n_\mu)/r, \quad (24b)$$

since the term of $O(1/r^2)$, being a sum of derivatives with respect to t', can have a zero time-average. If $G^{(2)}{}_{\mu t}$ is of $O(K^2/r^2)$ and has a nonzero time-average, consistency can be achieved only if another term of time-average $O(K^2/r^2)$ at vacuum be added to the source of (6). Note that there is no plane-wave solution for (6') [74].

It will be shown that there is no dynamic solution for (1) with a massive source. Let us define

$$\gamma_{\mu\nu} = \gamma^{(1)}{}_{\mu\nu} + \gamma^{(2)}{}_{\mu\nu} \,; \qquad \bar{\gamma}^{(i)}{}_{\mu\nu} = \gamma^{(i)}{}_{\mu\nu} - \frac{1}{2}$$

$$\eta_{\mu\nu} (\gamma^{(i)}{}_{cd} \eta^{cd}), \quad \text{where } i = 1, 2 \,;$$

and

$$\frac{1}{2}\partial^\alpha\partial_\alpha \bar{\gamma}^{(1)}{}_{\mu\nu} = - K \, T(m)_{\mu\nu} \,. \quad (25)$$

Then $\bar{\gamma}^{(1)}{}_{\mu\nu}$ is of a first-order; and $\gamma^{(2)}{}_{\mu\nu}$ is finite. On the other hand, from (6), one has

$$\frac{1}{2}\partial^\alpha\partial_\alpha \bar{\gamma}^{(2)}{}_{\mu\nu} + H^{(1)}{}_{\mu\nu} + G^{(2)}{}_{\mu\nu} = 0 \,. \quad (26)$$

Note that, for a dynamic case, equation (26) may not be satisfied. If (25) is a first-order approximation, $G^{(2)}{}_{\mu\nu}$ has a nonzero time-average of $O(K^2/r^2)$ [1]; and thus $\bar{\gamma}^{(2)}{}_{\mu\nu}$ cannot have a solution.

However, if $\bar{\gamma}^{(2)}{}_{\mu\nu}$ is also of the first-order of K, one cannot estimate $G^{(2)}{}_{\mu\nu}$ by assuming that $\bar{\gamma}^{(1)}{}_{\mu\nu}$ provides a first-order approximation. For example, (6) does not provide the first approximation for the static Schwarzschild solution, although it can be transformed to a form such that (6) provides a first-order approximation [8]. According to eq.(15), $\bar{\gamma}^{(2)}{}_{\mu\nu}$ will be a second order term if the sum $H^{(1)}{}_{\mu\nu}$ is of second order. From (22b), this would require $\partial^\mu \bar{\gamma}_{\mu\nu}$ being of second order. For weak gravity, it is known that a coordinate transformation would turn $\partial^\mu \bar{\gamma}_{\mu\nu}$ to a second order term [77, 87, 97]. (Eq. [26] implies that

$\partial^c \partial_c \gamma^{(2)}{}_{\mu\nu} - \partial^c [\partial_\nu \bar{\gamma}_{\mu c} + \partial_\mu \bar{\gamma}_{\nu c}]$ would be of second order.) Thus, it is possible to turn (25) to become an equation for a first-order approximation for weak gravity.

Physically, since it has been proven that (21) necessarily gives a first-order approximation [8], a failure of such a coordinate transformation means only that such a solution is not valid in physics. Moreover, for the dynamic of massive matter, experiment [95] supports the fact that Maxwell-Newton Approximation (21) is related to a dynamic solution of weak gravity [7]. Otherwise, not only is Einstein's radiation formula not valid, but the theoretical framework of general relativity should be re-examined. In other words, theoretical considerations in physics as well as experiments eliminate other unverified speculations thought to be possible since 1957.

As shown, the difficulty comes from the assumption of boundedness, which allows the existence of a bounded first-order approximation, which in turn implies that a time-average of the radiative part of $G^{(2)}{}_{\mu\nu}$ is non-zero. The present method has an advantage over Fock's approach to obtaining logarithmic divergence [7, 87] for being simple and clear.

In short, according to Einstein's radiation formula, a time average of $G^{(2)}{}_{\mu t}$ is non-zero and of $O(K^2/r^2)$ [7]. Although (21) implies $G^{(1)}{}_{\mu t}$ is of order K^2, its terms of $O(1/r^2)$ can have a zero time average because $G^{(1)}{}_{\mu t}$ is linear on the metric elements. Thus, (6') cannot be satisfied. Nevertheless, a static metric can satisfy (6), since both $G^{(1)}{}_{\mu\nu}$ and $G^{(2)}{}_{\mu\nu}$ are of $O(K^2/r^4)$ in vacuum. Thus, that a gravitational wave carries energy-momentum does not follow from the fact that $G^{(2)}{}_{\mu\nu}$ can be identified with a gravitational energy-stress [1, 77]. Note that $G_{\mu t} = -KT(m)_{\mu t}$ are constraints on the initial data.

In conclusion, in disagreement with the physical requirement, assuming the existence of dynamic solutions of weak gravity for (6) [51, 55, 87, 96-101] is invalid. This means that the calculations [45, 46] on the binary pulsar experiments should, in principle, be re-

addressed [24]. This explains also that an attempt by Christodoulou and Klainerman [81] to construct bounded "dynamic" solutions for $G_{\mu\nu} = 0$ fails to relate to a dynamic source and to be compatible with (21) [80].

7. The Anti-Gravity Coupling and Invalidity of the Space-time Singularity Theorems to Physics

From the above analysis, there is a conflict between the Einstein equation, which has no dynamic solution and its linearized equation, which has a dynamic solution. The conflict is due to that the second order terms $G^{(2)}{}_{\mu\nu}$ cannot be eliminated in the Non-linear Einstein equation. Thus, a simple solution is the 1995 update of the Einstein equation [7] as follows:

$$G_{\mu\nu} \equiv R_{\mu\nu} - \frac{1}{2} g_{\mu\nu} R = -K [T(m)_{\mu\nu} - t(g)_{\mu\nu}], \quad (27)$$

where $t(g)_{\mu\nu}$ is the energy-stress tensors for gravity. Then,

$$\nabla^\mu T(m)_{\mu\nu} = 0, \quad \text{and} \quad \nabla^\mu t(g)_{\mu\nu} = 0. \quad (28)$$

Equation (28) implies that the equivalence principle would be satisfied. From (27), the equation in vacuum is

$$G_{\mu\nu} \equiv R_{\mu\nu} - \frac{1}{2} g_{\mu\nu} R = K t(g)_{\mu\nu}. \quad (27')$$

Note that $t(g)_{\mu\nu}$ is equivalent to $G^{(2)}{}_{\mu\nu}$ (and Einstein's gravitational pseudotensor) in terms of his radiation formula. The fact that $t(g)_{\mu\nu}$ and $G^{(2)}{}_{\mu\nu}$ are related under some circumstances does not cause $G^{(2)}{}_{\mu\nu}$ to be an energy-stress nor $t(g)_{\mu\nu}$ a geometric part, just as $G_{\mu\nu}$ and $T_{\mu\nu}$ must be considered as distinct in the Einstein equation (6).

When gravitational wave is present, the gravitational energy-stress tensor $t(g)_{\mu\nu}$ is non-zero. Thus, a radiation does carry energy-momentum as physics requires. This explains also that the absence of an anti-gravity coupling which is determined by Einstein's radiation formula, is the physical reason that the 1915 Einstein equation (1) is incompatible with radiation.

Note that the radiation of the binary pulsar can be calculated without detailed knowledge of $t(g)_{\mu\nu}$. From (27'), the approximate value of $t(g)_{\mu\nu}$ at vacuum can be calculated through $G_{\mu\nu}/K$ as before since the first-order approximation of $g_{\mu\nu}$ can be calculated through (21). In view of the facts that $Kt(g)_{\mu\nu}$ is of the fifth order in a post-Newtonian approximation, that the deceleration due to radiation is of the three and a half order in a post-Newtonian approximation [77] and that the perihelion of

Mercury was successfully calculated with the second-order approximation from (6), the orbits of the binary pulsar can be calculated with the second-order post-Newtonian approximation of (27) by using (6). Thus, the calculation approaches of Damour and Taylor [102, 103] would be essentially valid except that they did not realize the crucial fact that (21) is actually an approximation of the updated equation (27) [86].

In light of the above, the Hulse-Taylor experiments support the anti-gravity coupling being crucial to the existence of the gravitational wave [7, 47], and (21) being an approximation of weak waves generated by massive matter. Thus, it has been experimentally verified that Einstein equation (6) is not compatible with radiation, but the updated Einstein equation does.

The 1995 updated Einstein equation actually was first proposed by Lorentz [104] and Levi-Civita [105] in the following form,

$$\kappa t(g)_{ab} = G_{ab} + \kappa T_{ab} \qquad (29)$$

where $t(g)_{ab}$ is the gravitational energy-stress tensor, G_{ab} is the Einstein tensor, and T_{ab} is the sum of other massive energy-stress tensors. Then, the gravitational energy-stress tensor takes a covariant form, although they have not proved its necessity. However, Einstein [106] objected to this form on the grounds that his field equation implies $t(g)_{ab} = 0$. Now, Einstein is clearly wrong since his equation is proven invalid for the dynamic case. Thus, eq. (27) should be called the Lorentz-Levi-Einstein equation.

One might object that while eq. (27) is consistent with the linearized equation for the massive case and can do an approximate calculation for the gravitational radiation, it is still not clear that it is the exact equation. For this, our position is that this is the best we can get so far. Further verification can be done only after the exact form of the gravitational energy-stress tensor $t(g)_{ab}$ is known. Moreover, if the unique sign for couplings could be attributed to a general $E = mc^2$,[20] the non-unique signs of coupling would suggest that $E = mc^2$ is only conditionally valid. We shall show explicitly that the electromagnetic energy is not equivalent to mass.

It should be noted also that the anti-gravity coupling would appear in where the gravitational wave is present. For instance, it is necessary to appear in the equation for the calculation on the gravitational waves generated by an electromagnetic wave. For the validity of the calculation on light bending, it is necessary that an electromagnetic wave would generate a negligible gravitational wave because this was implicitly assumed in such a calculation. For this case the related equation is the following:

$$G_{ab} = -K[T(E)_{ab} - T(p)_{ab}], \quad \text{and}$$

$$T_{ab} = -T(g)_{ab} = T(E)_{ab} - T(P)_{ab}, \qquad (30)$$

where $T(E)_{ab}$ and $T(P)_{ab}$ are the energy-stress tensors for the electromagnetic wave and the related photons. The presence of the photonic energy-stress is necessary; otherwise there is no bounded gravitational wave solution for equation (30) [107]. Thus, the anti-gravity coupling must be present for any dynamic case.

In Einstein's initial assumption, the photons consist of only electromagnetic energy because general relativity has not been conceived. If the photons consist of only electromagnetic energy, there is a conflict since the photonic energy can be equivalent to mass and the electromagnetic energy-stress tensor is traceless. Now, this conflict is resolved since the photonic energy is the sum of electro-magnetic energy and gravitational energy. Both quantum theory and relativity are based on the phenomena of light. The gravity of photons shows that there is a link between them. It is gravity that makes the notion of photons compatible with electromagnetic waves. Einstein probably would smile heartily since his formula confirms the link that relates gravity to quantum theory.

It should be noted that the existence of an anti-gravity coupling means the energy conditions in the singularity theorems [2] are not valid for a dynamic situation. Thus, the existence of singularity is not certain, and the claim of inevitably breaking of general relativity is baseless since these singularity theorems have been proven to be unrealistic in physics.

8. Errors of the Mathematicians, such as Atiyah, Penrose, Witten, and Yau

While physicists can make errors because of inadequate background in mathematics, one may expect that professional mathematician would help improve the situation. However, this expectation may not always be fulfilled. The reason is that a mathematician may not understand the physics involved and thus it may not be helpful, but also could make the situation worse. A good example is the cooperation between Einstein and Grossman. They even wrote a joint paper, but from which it is clear that they did not understand each other.

However, the situation could be worse if they had misled each other.

A bad example is the participation in physics by mathematician Roger Penrose. He was misled by invalid understanding of the formula that $E = mc^2$ is always valid and thus he believed that all the coupling of energy-momentum tensors must have the same sign. Then his talent comes up with the space-time singularity theorems without realizing such an assumption is invalid in physics. Then, he and Hawking convince the physicists that such singularities must exist and general relativity is not valid in the microscopic scale. Thus, an accompanying error is that physicists failed to see that the photons must include gravitational components since photons can be converted into mass [107, 108]. This is supported by the fact that the meson π_0 can decay into two γ rays. However, the electromagnetic energy alone cannot be equivalent to mass because the trace of the massive energy-momentum tensor is non-zero.

Moreover, the singularity theorems show only the breaking down of theories of the Wheeler-Hawking school, which are actually different from general relativity. Their theories differ from general relativity in at least the following important aspects:

1) They reject an anti-gravity coupling, which is considered as necessary to obtain a bounded dynamic solution.

2) They implicitly replaced Einstein's equivalence principle with merely the mathematical requirement of the existence of local Minkowski spaces [7, 61].

3) They do not consider physical principles such as the principle of causality of which the satisfaction is vital for a physical solution such that Einstein's equivalence principle can be applicable.

Thus, in spite of declaring their theories as the development of general relativity [1], these theories actually contradict crucial features that are indispensable in Einstein's theory of general relativity. More importantly, in the development of their so-called "orthodox theory," they violate physical principles that took generations to establish. As a result, Einstein's theory has been unfairly considered as irrelevant in the eyes of many physicists. They also support their accusations with false evidence [20, 29].

Moreover, most of the mathematicians were not aware of the need of a term with anti-gravity coupling such that a bounded dynamic solution would be obtained. Otherwise there is no bounded dynamic solution for the Einstein equation. They naively thought without adequate deliberation, that a dynamic solution which is asymptotically flat would be the most normal physical solution. This leads to the positive energy theorem of Schoen and Yau [68]. From the free encyclopedia Wikipedia, the contributions of Prof. Yau were naively summarized as follows:

"Yau's contributions have had a significant impact on both physics and mathematics. Calabi–Yau manifolds are among the 'standard tool kit' for string theorists today. He has been active at the interface between geometry and theoretical physics. His proof of the positive energy theorem in general relativity demonstrated, sixty years after its discovery, that Einstein's theory is consistent and stable. His proof of the Calabi conjecture allowed physicists, using Calabi-Yau compactification, to show that string theory is a viable candidate for a unified theory of nature."

Thus, it was claimed that Yau's "proof " of the positive energy theorem [68, 69] in general relativity would have profound influence that leads to even the large research efforts on string theory. Based on his proof, it was claimed that Einstein's theory is consistent and stable. This would be in a direct conflict with the fact the there is no dynamic solution for the Einstein equation and would not be supported by the solutions in the literature [55].

Now, let us examine their theorem that would imply that flat space-time is stable, a fundamental issue for the theory of general relativity. Briefly, the positive mass conjecture says that if a three-dimensional manifold has positive scalar curvature and is asymptotically flat, then a constant that appears in the asymptotic expansion of the metric is positive. A crucial assumption in the theorem of Schoen and Yau is that the solution is asymptotically flat. However, since the Einstein equation has no dynamic solution, which is bounded, the assumption of asymptotically flat implies that the solution is a stable solution such as the Schwarzschild solution, the harmonic solution, the Kerr solution, etc.

Therefore, Schoen and Yau actually prove a trivial result that the total mass of a stable solution is positive. Note that since the dynamic case is actually excluded from the consideration in the positive energy theorem, this explains why it was found from such a theorem that Einstein's theory is consistent and stable. This is, of course, misleading.

Due to inadequacy in pure mathematics among physicists, the non-existence of dynamic solutions for the Einstein equation was not recognized by physicists. So, Yau could only invalidly assume the existence of a bounded dynamic solution. Thus, the positive energy theorem of Schoen and Yau also continues such an error. Although Yau may not have made errors in mathematics, their positive energy theorem produced not only just useless but also misleading results in physics. Yau failed to see this problem of misleading since he has not attempted to find explicit examples to illustrate their theorem. Nevertheless, in awarding him a Fields Medal in 1982, this theorem is cited as an achievement. Moreover, E. Witten made the same mistake in his alternative proof [69], but that was also cited as an achievement for his Fields Medal in 1990.[21] Apparently, Atiyah also failed to understand this issue.[22]

In fact, Yau [68], Witten [69], and Christodoulou [81] make essentially the same error of defining a set of solutions that actually includes no dynamic solutions. Their fatal error is that they neglected to find explicit examples to support their claims. Had they tried, they should have discovered their errors. Note that Yau has wisely avoided committing himself to the errors of Christodoulou & Klainerman, by claiming that his earlier interest has changed [81]. However, he was unable to see that the binary pulsars experiment of Hulse & Taylor not only confirms that there is no dynamic solution but also that the signs of coupling constants are not unique [7-9]. In fact, Yau has made the same errors of Penrose and Hawking [2], and implicitly uses the invalid assumption of unique sign in his positive energy theorem of 1981. Nevertheless, Prof. Yau is a good mathematician as shown by his other works although he does not understand physics well. Since the Einstein equation must be modified for a dynamic case, their positive energy theorem is also irrelevant to physics just as the space-time singularity theorem of Penrose and Hawking.

Note that Einstein & Rosen [109] are the first who recognized the non-existence of wave solution for the Einstein equation. The facts that Atiyah [82], Hilbert [110], Witten [69], and Yau [68] were unable (or

neglected) to identify their errors, would have misleadingly created a false impression that Einstein, the Wheeler School, and their associates did not make errors in mathematics.

9. E = mc², the Reissner-Nordstrom Metric, and the Question of Black Holes

The existence of the anti-gravity coupling raised a question whether the formula $E = mc^2$ is unconditionally valid. It is found that this is only a speculation that Einstein failed to prove its general validity (1905-1909) [111]. Moreover, since the electromagnetic energy-stress tensor is traceless, an electromagnetic energy-stress tensor would generate gravitation which does not change the Ricci curvature R in the Einstein equation. Note that if unverified speculation $E = mc^2$ being unconditionally valid, this would imply that attractive gravity would always increase as the energy of sources increase. And this is the central implicit assumption for the existence of a black hole [20]. The fact that a black hole has never been observed would imply that some energy can generate repulsive gravitation. It seems that the Riessner-Nordstrom metric [1] for a charged particle would answer the above issues.[23]

Now, let us examine the Reissner-Nordstrom metric [1] (with c =1) as follows:

$$ds^2 = \left(1 - \frac{2M}{r} + \frac{q^2}{r^2}\right)dt^2 - \left(1 - \frac{2M}{r} + \frac{q^2}{r^2}\right)^{-1} dr^2 - r^2 d\Omega^2 \quad (31)$$

where q and M are the charge and mass of a particle, and r is the radial distance (in terms of the Euclidean-like structure [54]) from the particle center. In this metric (31), the gravitational components generated by electricity have not only a very different radial coordinate dependence but also a different sign that makes it a new repulsive gravity in general relativity.

Nevertheless, some argued that the effective mass could be considered as

$$M - \frac{q^2}{2r}, \quad (32)$$

because the total electric energy outside a sphere of radius r is $q^2/2r$ [112, 113], and thus (32) could be interpreted as supporting $m = E/c^2$ at least for electromagnetic energy. If the electric energy has a mass equivalence, an increase of such energy should lead to an increment of gravitational strength. However, the strength of a gravitational force, from metric (31), decreases everywhere.

Moreover, the gravitational forces would be different from the force created by the "effective mass" $M - q^2/2r$ because

$$-\frac{1}{2}\frac{\partial}{\partial r}\left(1-\frac{2M}{r}+\frac{q^2}{r^2}\right)=-\left(\frac{M}{r^2}-\frac{q^2}{r^3}\right)>-\frac{1}{r^2}\left(M-\frac{q^2}{2r}\right) \quad (33).$$

Thus Will was defeated because he could not defend his interpretation of $m = E/c^2$ [28].

The general validity of $E = mc^2$ was questioned because the binary pulsars experiment that the coupling constants necessarily have different signs [7-9]. Nevertheless, with supports from editorials of Nature, the Physical Review D, and Science, Will continued to misinterpret [6] the formula. Also, some theorists [114, 115] incorrectly argued that M in (1) includes the external electric energy. To see this error clearly, one has to go through the derivation of the Reissner-Nordstrom metric.

9.1. *Derivation the Reissner-Nordstrom Metric and the Repulsive Charge-Mass Interaction*

It seems that mass M in (1) as a "total mass" that includes the electric energy, would be allowed if you are careless. However, a close examination shows that this is invalid.

According to Einstein, the static field equation for the metric is [5],

$$G_{\mu\nu} \equiv R_{\mu\nu} - \frac{1}{2}g_{\mu\nu}R = -8\pi\, T_{\mu\nu},$$
$$(34)$$

or

$$R_{\mu\nu} = -8\pi[T_{\mu\nu} - \frac{1}{2}g_{\mu\nu}T], \quad \text{where} \quad T = g^{\alpha\beta}T_{\alpha\beta}$$

In this equation, the energy stress tensor $T_{\mu\nu}$ is the sum of any type of energy-stress tensor. For the Reissner-Nordstrom metric, it includes at least the massive energy-stress tensor and the electromagnetic energy-stress tensor. They differ by that the electromagnetic energy-stress tensor is traceless whereas the massive energy-stress tensor is not.

If one assumes that the metric has the following form,

$$ds^2 = f\,dt^2 - h\,dr^2 - r^2(d\theta^2 + \sin^2\theta\,d\phi^2) \quad (35)$$

then, as shown by Wald [2], at the region outside the particle ($r > r_0$) we have

$$-R_{00}=\frac{1}{2}(fh)^{-1/2}\frac{d}{dr}\left[(fh)^{-1/2}f'\right]+(fhr)^{-1}f' \quad (36a)$$

$$-R_{11}=-\frac{1}{2}(fh)^{-1/2}\frac{d}{dr}\left[(fh)^{-1/2}f'\right]+(h^2r)^{-1}h', \quad (36b)$$

$$-R_{22}=-\frac{1}{2}(rfh)^{-1}f'+\frac{1}{2}(h^2r)^{-1}h'+r^{-2}(1-h^{-1}) \quad (36c)$$

Moreover, outside the particle we have

$$T(m)_{\mu\nu} = 0 \text{ for } \qquad r > r_0 \quad (37a)$$

But

$$T(m)_{00} = \rho(r),\ T(m)_{11} = T(m)_{22} = T(m)_{33} = P(r),$$

when $\qquad r < r_0 \qquad (37b)$

where $P(r)$ is the pressure of the perfect fluid model.

Because the electric energy-stress tensor $T(E)_{\mu\nu}$ is traceless, we also have, for $r > r_0$,

$$R_{00} = -R_{11} = R_{22} = -E^2, \quad \text{where} \quad \overline{E} = \frac{q}{r^3}\overline{r} \quad (38)$$

is the electric field, according to Misner et al. [1; p. 841]. If $h = 1/f$ in metric (35), then (36) is reduced to

$$-R_{00} = R_{11} = \frac{1}{2}f''+r^{-1}f' = E^2 \quad (39a)$$

and

$$-R_{22} = -r^{-1}f'+r^{-2}(1-f) = E^2 \quad (39b)$$

Moreover, if $f = \left(1 - \frac{2M}{r} + \frac{q^2}{r^2}\right)$ as in metric (31), then

we have, in consistent with (38),

$$\frac{q^2}{r^2} = r^2 E^2 \quad (40)$$

Thus, from the above derivation, it seems there is no restriction on the mass M of metric (35). However, from (38), it is clear that M in metric (35) cannot include the electric energy (outside the particle) because it has been represented in (38).

9.2. *Misinterpretations of the Reissner-Nordstrom Metric and Related Experiments*

Nevertheless, Herrera, Santos, & Skea [115], argued that M in (1) involves the electric energy. They follow the error of Whittaker [116] and Tolman [117] who believed the equivalence of mass and electric energy. They defined the active gravitational mass density μ with the electromagnetic energy tensor E^α_β as $\mu = E^0_0 - E^i_i$ and the active mass in a volume V_a is given by

$$m_a(r) = \int_{V_a} \mu(-g)^{1/2}\,dx^1 dx^2 dx^3, \quad (41)$$

where g is the four dimensional determinant of the metric. It thus follows that, for a particle with charge Q, one has

$$m_a(\infty) - m_a(r) = \int_r^\infty \frac{Q^2}{r^2} dr \text{ , and } \quad m_a(r) = M - \frac{Q^2}{r},$$

$$\text{where } \quad m_a(\infty) = M \qquad (42)$$

Thus $m_a(r)$ would be in agreement with that the total force is proportion to

$$\frac{1}{2} \frac{\partial}{\partial r} \left(1 - \frac{2M}{r} + \frac{Q^2}{r^2} \right) = (M - \frac{Q^2}{r}) \frac{1}{r^2}$$

$$= \left(m_a(r_0) + Q^2 (\frac{1}{r_0} - \frac{1}{r}) \right) \frac{1}{r^2} \qquad (43a)$$

because $M = m_a(r) + Q^2 / r = \left(m_a(r_0) + Q^2 / r_0 \right),$ (43b)

where r_0 is the radius of the particle. However, (42) does not agree with (32) because

$$-2(M - \frac{Q^2}{r}) \frac{1}{r} \neq \left(-\frac{2M}{r} + \frac{Q^2}{r^2} \right) \qquad (43c)$$

Eq. (43a) implies that the weight of a charged metal ball would increase when the charge Q is increased. According to eq. (41), $m_a(r_0)$ would increase as the charge Q increases. Thus a test of their interpretation is whether the repulsive effect can be detected.

The above approach is essentially the same as that of Pekeris [114], who gets a similar metric in 1982 as follows:

$$ds^2 = e^\nu dt^2 - e^{-\nu} dR^2 - R^2 d\Omega^2 \text{ where } R^3 = r^3 + r_0 \text{ (44a)}$$

$$e^\nu = \left(1 - \frac{2M_{mat}}{R} - \frac{2M_{em}}{R} + \frac{Q^2}{R^2} \right) = \left(1 - \frac{2M}{R} + \frac{Q^2}{R^2} \right),$$

where $M_{em} = Q^2/r_0$, and $M = M_{mat} + M_{em}$ (44b)

The difference is due to that Pekeris requires that $|g_{\mu\nu}| = g = -1$. Thus, the approach of Herrera et al. [115] is essentially what Pekeris had done. Apparently, theorists have run out of ways that can be used against the repulsive force. Although the Riessner-Nordstrom metric and the other two metrics look the same, they are different because the mass M means different in respective metrics.

Nevertheless, Herrera et al. [115] are not alone on this error. They have plenty of company who make the same mistake. A better known one is the Nobel Laureate, 't Hooft who even claimed that the electric energy of an electron contributed to the inertial mass of an electron [61]. It is such popular confusion among physicists that delays the discovery of the charge-mass interaction.

9.3 *The Invalidity of E = mc² for the Electric Energy and the Question of the Black Holes*

However, if M includes energy outside the particle as shown in (43b), this is in conflict with (38), and thus their derivation is actually invalid. On the other hand, if the mass M is just the inertial mass of the particle, the weight of a charged metal ball can be reduced [118]. Thus, as Lo expected [6], experiments of Tsipenyuk and Andreev on two metal balls [119] rejects eq. (42), and their misinterpretations since the charged ball has reduced weight. This is an experimental direct proof that the electric energy is not equivalent to mass since they generate different gravities; one is attractive to mass and the other is repulsive to mass. We recommend that the details of such experiments [118] should be continued such that this static case of general relativity is fully verified.

Note that the appearing of the repulsive gravitation is important because it would solve a puzzle as to why we have never seen a black hole. If gravity is always attractive to mass, Wheeler simulation convinces him that a black hole must be formed [20]. His implicit assumption is that gravity is always attractive, but such an assumption has been proven incorrect. Another piece of information that led to Wheeler's belief of the existence of black holes is the existence of space-time singularities, the mathematical theorems proven by Hawking and Penrose [2]. Now, because the necessity of the existence of the anti-gravity coupling the energy conditions of their theorems cannot be satisfied. Thus, their space-time singularity theorems are actually irrelevant to physics. Moreover, now the Einstein equation is known to be invalid for the dynamic case [7, 67], thus the derivation for the existence of the black hole is not valid. In short, if theorists still believe the existence of black holes in spite of these, they must find new justifications for their claim.

More important, this new repulsive force is crucial for establishing the unification of gravitation and electromagnetism.

10. The Charge-Mass Repulsive Force and Einstein's Unification.

To show the static repulsive effect, one needs to consider only g_{tt} in metric (31). According to Einstein [6, 3],

$$\frac{d^2 x^\mu}{ds^2} + \Gamma^\mu{}_{\alpha\beta} \frac{dx^\mu}{ds} \frac{dx^\nu}{ds} = 0, \qquad (45)$$

where $\Gamma^\mu{}_{\alpha\beta} = (\partial_\alpha g_{\nu\beta} + \partial_\beta g_{\nu\alpha} - \partial_\nu g_{\alpha\beta}) g^{\mu\nu} / 2$

and $ds^2 = g_{\mu\nu} dx^\mu dx^\nu$. Let us consider only the static case. (One need not worry whether the gauge is physically valid because the gauge affects only the second order approximation of g_{tt} [77].) For a particle P with mass m at **r**, the force on P is

$$-m \frac{M}{r^2} + m \frac{q^2}{r^3} \qquad (46)$$

in the first order approximation because $g^{rr} \cong -1$. Thus, the second term is a repulsive force.

If the particles are at rest, then the force acting on the charged particle Q has the same magnitude

$$(m \frac{M}{r^2} - m \frac{q^2}{r^3}) \hat{r}, \text{where} \quad \hat{r} \text{ is a unit vector} \qquad (47)$$

because the action and reaction forces are equal and in the opposite directions. However, for the motion of the charged particle with mass M, if one calculates the metric according to the particle P of mass m, only the first term is obtained.

Then, it is necessary to have a repulsive force with the coupling q^2 to the charged particle Q in a gravitational field generated by masses. Thus, force (47) to particle Q is beyond current theoretical framework of gravitation + electromagnetism. In other words, as predicted by Lo, Goldstein, & Napier [46], general relativity leads to a realization of the inadequacy of general relativity.

The charge-mass repulsive force for two point-like particles is inversely proportional to the cube power (instead of the square) of the distances r between the two particles with mass m and charge q respectively, i.e.,

$$mq^2/r^3 \qquad (48)$$

This means that such a repulsive force would become weak faster than gravity at long distance, although it would be much stronger at very small distance. Moreover, this force is proportional to the square of the charge q, and thus is independent of the charge sign. Such characteristics would make the repulsive effects verifiable [118] because a concentration of electrons would increase such repulsion.

The term of repulsive force in (31) comes from the electric energy [5]. An immediate question would be whether such a charge-mass repulsive force mq^2/r^3 is subjected to electromagnetic screening. It is conjectured that this force, being independent of a charge sign, would not be subjected to such a screening although it should be according to general relativity. From the viewpoint of physics in a field theory, this force can be considered as a result of a field created by the mass m and the field interacts with the q^2. Thus such a field is independent of the electromagnetic field and is beyond general relativity. Thus, the need of unification is established.

11. Extension of Einstein's Theory and the Five-Dimensional Relativity

If we consider the coupling with q^2, this naturally leads to a five dimensional space of Lo et al. [46]. This started with Kaluza's [120] five-dimensional general relativity, and he maintains the equation of motion as being a geodesic equation. Based on the cylindrical condition [24] that reduces the five variables to four, this theory reproduces the Einstein equation and the Maxwell equation if the "extra" metric elements are considered as constant or negligible. Einstein and Pauli [13] also wrote a paper to continue the work of Kaluza. The five-dimensional relativity does have the coupling with the square of a charge if the "extra" metric elements are retained. If cylindrical condition is not imposed, the radiation reaction force would also be accounted for [46].

Now let us give an description of the five-dimension-al theory. The five dimensional geodesic of a particle is

$$\frac{d}{ds}\left(g_{ik} \frac{dx^k}{ds}\right) = \frac{1}{2} \frac{\partial g_{kl}}{\partial x^i} \frac{dx^k}{ds} \frac{dx^l}{ds} + \left(\frac{\partial g_{5k}}{\partial x^i} - \frac{\partial g_{5i}}{\partial x^k}\right) \frac{dx^5}{ds} \frac{dx^k}{ds} - \Gamma_{i,55} \frac{dx^5}{ds} \frac{dx^5}{ds} - g_{i5} \frac{d^2 x^5}{ds^2}, \qquad (49a)$$

$$\frac{d}{ds}\left(g_{5k} \frac{dx^k}{ds} + \frac{1}{2} g_{55} \frac{dx^5}{ds}\right) = \Gamma_{k,55} \frac{dx^5}{ds} \frac{dx^k}{ds} - \frac{1}{2} g_{55} \frac{d^2 x^5}{ds^2} + \frac{1}{2} \frac{\partial g_{kl}}{\partial x^5} \frac{dx^l}{ds} \frac{dx^k}{ds}, \qquad (49b)$$

where $ds^2 = g_{\mu\nu}dx^\mu dx^\nu$, $\mu, \nu = 0, 1, 2, 3, 5$ ($d\tau^2 = g_{kl}dx^k dx^l$; $k, l = 0, 1, 2, 3$).

If instead of ds, $d\tau$ is used in (49), for a particle with charge q and mass M, the Lorentz force suggests

$$\frac{q}{Mc^2}\left(\frac{\partial A_i}{\partial x^k} - \frac{\partial A_k}{\partial x^i}\right) = \left(\frac{\partial g_{i5}}{\partial x^k} - \frac{\partial g_{k5}}{\partial x^i}\right)\frac{dx^5}{d\tau} \qquad (50a)$$

Thus

$$\frac{dx^5}{d\tau} = \frac{q}{Mc^2}\frac{1}{K}\; K\left(\frac{\partial A_i}{\partial x^k} - \frac{\partial A_k}{\partial x^i}\right) = \left(\frac{\partial g_{i5}}{\partial x^k} - \frac{\partial g_{k5}}{\partial x^i}\right) \quad \text{and} \quad \frac{d^2 x^5}{d\tau^2} = 0 \qquad (50b)$$

where K is a constant. It thus follows that (48) is reduced to

$$\frac{d}{d\tau}\left(g_{ik}\frac{dx^k}{d\tau}\right) = \frac{1}{2}\frac{\partial g_{kl}}{\partial x^i}\frac{dx^k}{d\tau}\frac{dx^l}{d\tau} + \left(\frac{\partial A_k}{\partial x^i} - \frac{\partial A_i}{\partial x^k}\right)\frac{q}{Mc^2}\frac{dx^k}{d\tau} - \Gamma_{i,55}\left(\frac{q}{Mc^2}\right)^2\frac{1}{K^2} \qquad (51a)$$

$$\frac{d}{d\tau}\left(g_{5k}\frac{dx^k}{d\tau} + \frac{1}{2}g_{55}\frac{q}{KMc^2}\right) = \Gamma_{k,55}\frac{q}{KMc^2}\frac{dx^k}{d\tau} + \frac{1}{2}\frac{\partial g_{kl}}{\partial x^5}\frac{dx^l}{d\tau}\frac{dx^k}{d\tau}. \qquad (51b)$$

One may ask what the physical meaning of the fifth dimension is. Our position is that the physical meaning of the fifth dimension is not yet very clear [46], except some physical meaning is given in the equation, $dx^5/d\tau = q/Mc^2K$ where M and q are respectively the mass and charge of a test particle, and K is a constant. This equation relates the fifth variable x^5 to τ.

The fifth dimension is assumed [46] as part of the physical reality, and the metric signature is $(+, -, -, -, -)$. We shall denote the fifth axis as the w-axis (w stands for "wunderbar", in memorial of Kaluza), and thus the coordinates are (t, w, x, y, z). Our approach is to find out the full physical meaning of the w-axis as our understanding gets deeper. In Physics things are often not defined clearly right at the beginning. For example, it takes us a long time to understand the energy-momentum conservation.

For a static case, we have the forces on the charged particle Q in the ρ-direction

$$-\frac{mM}{\rho^2} \approx \frac{Mc^2}{2}\frac{\partial g_{tt}}{\partial \rho}\frac{dct}{d\tau}\frac{dct}{d\tau}g^{\rho\rho},$$

$$\frac{mq^2}{\rho^3} \approx -\Gamma_{\rho,55}\frac{1}{K^2}\frac{q^2}{Mc^2}g^{\rho\rho} \qquad (52a)$$

and $\qquad \Gamma_{k,55}\dfrac{q}{KMc^2}\dfrac{dx^k}{d\tau} = 0,$ $\qquad$ where

$$\Gamma_{k,55} \equiv \frac{\partial g_{k5}}{\partial x^5} - \frac{1}{2}\frac{\partial g_{55}}{\partial x^k} = -\frac{1}{2}\frac{\partial g_{55}}{\partial x^k} \qquad (52b)$$

in the $(-r)$-direction. The meaning of (51b) is the energy momentum conservation. Thus,

$$g_{tt} = 1 - \frac{2m}{\rho c^2}, \qquad \text{and}$$

$$g_{55} = \frac{mMc^2}{\rho^2}K^2 + \text{constant。} \qquad (53)$$

In other words, g_{55} is a repulsive potential. Because g_{55} depends on M, it is a function of local property, and thus is difficult to calculate. This is different from the metric element g_{tt} that depends on a distant source of mass m.

On the other hand, because g_{55} is independent of q, $(\partial g_{55}/\partial \rho)/M$ depends only on the distant source m. Thus, this force, though acting on a charged particle, would penetrate electromagnetic screening. From the above, it is possible that a charge-mass repulsive potential would exist for a metric based on the mass M of the charged particle Q. However, because P is neutral, there is no charge-mass repulsion force (from $\Gamma_{k,55}$) on P.

Thus, general relativity must be extended to accommodate the charge-mass interaction, and five-dimensional relativity is a natural candidate to include such a force. According the five-dimensional theory of Lo et al. [46], the charge-mass interaction would penetrate a charged capacitor. On the other hand, we would not get a repulsive force acting on a test particle outside a capacitor from current four-dimensional theory. Since the electromagnetic field outside a capacitor would cancel out, there is no charge-mass interaction outside the capacitor. To verify the five-dimensional theory, one can simply test the repulsive force on a charged capacitor.

Einstein and Pauli [13], remarked in 1943

"When one tries to find a unified theory of
gravitational and electromagnetic fields, he

cannot help feeling that there is some truth in Kaluza's five-dimension theory."

It may turn out that their observation would be a prophecy for the future advancement of such unification.

12. Experimental Verification of the Mass-Charge Repulsive Force

The repulsive force in metric (31) can be detected with a neutral mass. To see this effect, one must have

$$\frac{1}{2}\frac{\partial}{\partial r}\left(1-\frac{2M}{r}+\frac{q^2}{r^2}\right)=\frac{M}{r^2}-\frac{q^2}{r^3}<0 \qquad (54)$$

Thus, repulsive gravity would be observed at $q^2/M > r$. For the electron the repulsive gravity would exist only inside the classical electron radius r_0 $(= 2.817\times 10^{-13}cm)$. Thus, it is very difficult to test a single charged particle.

However, for a charged metal ball with mass M and charge Q, the formula is similarly $0 > M/R^2 - Q^2/R^3$, where R is the distance from the center of the ball [118]. Thus, the attractive effect in gravity is proportional to mass related to the number of electrons, but the repulsive effect in gravity is proportional to square of charge related to the square of the number of electrons. Thus, when the electrons are sufficiently accumulated, the effect of repulsive gravity will be shown in a macroscopic distance.

Consider that Q and M consists of N electrons, i.e., $Q = Ne$, $M = Nm_e + M_0$, where M_0 is the mass of the metal ball, m_e and e are the mass and charge of an electron. To have sufficient electrons, the necessary condition is

$$N > \frac{R}{r_0}, \quad \text{where } r_0 = \frac{e^2}{m_e c^2} = 2.817\times 10^{-13}\text{ cm}. \qquad (55)$$

For example, if $R = 10$ cm, then it requires $N > 3.550\times 10^{13}$. Thus $Q = 5.683\times 10^{-7}$ Coulomb. Then, one would see the attractive and repulsive additional forces change hands. For this case, the repulsive force is

$$\frac{Q^2 m_p}{R^3} \qquad (56)$$

where m_p is the mass of the testing particle P. And the total force is

$$(\frac{M_0 + Nm_e}{R^2} - \frac{N^2 e^2}{R^3})m_p \qquad (57)$$

When condition (55) is satisfied for a certain R, the weight reduction due to the repulsive effect will be observed as the charge increases. The existence of this repulsive effect has already been verified by the weight reduction experiments of Tsipenyuk & Andrev [119]. However, since the repulsive force is very small, the interference of electricity could be comparatively large.

In the case of charged capacitor, the repulsive force should be proportional to the potential square, V^2 where V is the electric potential difference of the capacitor [121]. This has been verified by the experiments of Musha [122]. Thus, the factor of charge square in Eq. (48) is correct. It remains to verify the space dependence. In the Biefeld-Brown effect [123], the initial thrust is directional decided by the applied electric field. However, the weigh reduction effect is not directional and it stays if the potential does not change. This is verified by Liu [124], who measured the weight reduction with the roll-up capacitors.

13. The Current-Mass Interaction.

The magnetic energy would lead to an attractive force from a current toward a mass [20, 125]. Also, due to the fact that a charged capacitor has reduced weight, it is necessary to have the current-mass interaction to cancel out the charge-mass interaction. In other words, the existence of the current-mass attractive force would solve a puzzle, i.e., why a charged capacitor exhibits the charge-mass repulsive force since a charged capacitor has no additional electric charges? In a normal situation, the charge-mass repulsive force would be cancelled by other forms of the current-mass force as Galileo -- Newton and Einstein implicitly assumed. This general force is related to the static charge-mass repulsive force in a way similar to the Lorentz force is related to the Coulomb force.

In fact, the existence of such a current-mass attractive force has been verified by Martin Tajmar and Clovis de Matos [125] from the European Space Agency. It is found that a spinning ring of superconducting material increases its weight much more than expected, and they claimed that general relativity had been proven wrong. However, according to quantum theory, spinning super-conductors should produce a weak magnetic field. Thus, they are also measuring the interaction between an electric current and the earth, i.e. an effect of the current-mass interaction! The current-mass interaction would generate a force which is perpendicular to the current.

This characteristic explains why an alternative current on the capacitor would also make a capacitor reduce its weight as the case of charged capacitors [122]. The alternative current would create an attractive force parallel to the surface of a flat capacitor. However, such an interaction would not cancel the repulsive force that is perpendicular to the surface. It follows that, just as the case of a charged capacitor, there are repulsive forces in the perpendicular direction of the surface. Note that our explanation on the weight reduction of alternative current is very different from other theories [122] that treat the weight reduction as due to the alternative current.

One may ask what the formula for the current-mass force is? However, unlike the static charge-mass repulsive force, which can be derived from general relativity; this general force would be beyond general relativity since a current-mass interaction would involve the acceleration of a charge, this force would be time-dependent and generates electromagnetic radiation. Moreover, when the radiation is involved, the radiation reaction force and the variable of the fifth dimension must be considered [46]. Thus, we are not ready to derive the current-mass interaction yet.

Nevertheless, we may assume that, for a charged capacitor, the resulting force is the interaction of net macroscopic charges with the mass [126]. The irradiated ball has the extra electrons compared to a normal ball [119]. A spinning ring of superconducting material has the electric currents that are attractive to the earth [125]. This also explains a predicted phenomenon, which is also reported by Liu [124] that it takes time for a capacitor to recover its weight after being discharged [126]. This was observed by Liu because his rolled-up capacitors keep heat better. A discharged capacitor needs time to dissipate the heat generated by discharging, and the motion of its charges would accordingly recover to normal.

14. Charged Capacitors, the Gravitator, and the Charge-Mass Interaction

To avoid the interference effects, it would be desirable to screen the electromagnetic effects out. It is known that a charged capacitor reduces its weight. Initially, Thomas Townsend Brown [123] developed an electric capacitor device that he termed a gravitator because he believes that somehow the gravitational field becomes affected by the plate high voltage charge. When energized with up to 150,000 volts of direct current,

Brown's gravitator developed a thrust in the direction of its positively charged end. When oriented up right on a scale and energized, it proceeded to gain or lose that amount of weight depending upon the charge polarity. It become lighter when the positive end faced up and heaver when the negative end face up.

His efforts were later joined by Paul A. Biefeld, at Dennison University. This effect is now known as Biefeld-Brown (B-B) effect [123]. The B-B effect is often referred to as due to the electrogravitics, which is defined by Valone [123] as "electricity used to create a force that depends upon an object's mass". Apparently the B-B effect consists of two parts: 1) the initial thrust is due to the electric potential that moves the electrons to the positive post; and 2) the subsequent lift is due to the concentration of the positive and negative charges separately. Because if the second point. a capacitor would become a lifter, which is able to lift its own weight plus a payload after being charged with a high voltage (about 40 kilovolts), but without continuous supply of electric energy [123].

A lifter could get to work by charging the wire to either a positive or a negative potential. It has been determined that the thrust is not due to ion wind effects [123]. Thus, the lifting is generated by changing something of the lifter with one high voltage charge.

A problem of the B-B effect is that this effect cannot be explained satisfactory with current theories. For instance, according to Einstein's $E = mc^2$ [127], a charged capacitor should increase its weight, but it actually deceases its weight [10]. However, some theorists tried to work out a theory within the current theoretical framework with imaginative but inadequately justified assumptions, instead of working out a new law based on observations and/or with established theories [122]. Naturally, this did not get acceptance, and many even incorrectly regarded the B-B effect as just experimental errors. Perhaps, it is time to remind theorists that the current physical laws are originally obtained from and tested by observations. In this paper, we try to take this established approach.

In a charged capacitor, the number of charged particles does not change and the only change is the state of motion of some electrons that have become statically concentrated instead of moving normally in orbits; and such a concentration makes a repulsive force appear. Since such a force does not appear before the concentration, it is clear that such a force was cancelled out by the force created by the motion of the electrons. In other words, the repulsive force generated by the static protons and the static electrons were cancelled by the force generated by the motion of the initially moving electrons.

However, this repulsive force cannot be proportional to the charge although it is acting on a charged particle because of its charge. We have equal numbers of negatively charged particles (electrons) and positively charged particles (protons) and that would lead to the cancellation of the forces generated by particles of different charges. However, if such a force is proportional to the charge square, then these two kind of forces would be added up, instead of cancelled out. Moreover, since the lifter has a limited height [123], one should expect that this repulsive force would diminish faster than the gravitational force. Thus, if we assume that the force is proportional to mass, as usual, the static charge-mass interaction would be a repulsive force.

Such a repulsive force F_r between a particle with charge e and another particle of mass m would have the following form,

$$F_r = Kme^2/r^n \quad \text{where } n > 2 , \tag{58}$$

r is the distance between the two particles, and K is a coupling constant. In formula (58), the coupling constant K and n the power of r is determined by experiments. The simplest case is n = 3.

Formula (58) is derived from the observations with common physical sense. Thus, it can be used as guidance for related discussions in physics. Such an approach has to be taken because we recognized that there is little in common between the charge-mass interaction and current four-dimensional theories. This is different from the derivations with very imaginative assumptions without adequate justifications [122]. One may ask whether the lighter weight of a capacitor after charged could be due to a decrease of mass. Such a speculation is ruled out. Inside a capacitor the increased energy due to being charged would not be pure electromagnetic energy such that, for the total internal energy, Einstein's formula $E = mc^2$ is valid.

15. Conclusions and Discussions

In general relativity, the existence of gravitational wave, a signal of a dynamic solution, is a crucial test of the field equation. Thus, an important question is: what does the gravitational field of a radiating asymptotically Minkowskian system look like? Without experimental inputs, to answer this question would be very difficult. Bondi, vander Burg, & Metzner [129] commented, "it is never entirely clear whether solutions derived by the usual method of linear approximation necessarily correspond in every case to exact solutions, or whether there might be spurious linear solutions which are not in any sense approximations to exact ones." Thus, in calculating gravitational waves, problems were incorrectly considered as due to the method rather than inherent in the equations. Another serious problem is that theorists incorrectly reject a repulsive gravitation [121].

Physically, it is natural to continue assuming Einstein's notion of weak gravity is valid. (Boundedness, as a physical requirement, may not be compatible with a nonlinear field equation.) The complexity of the Einstein equation makes it very difficult to have a closed form. Thus, it is necessary that a method of expansion should be used to examine the problem of weak gravity.

A factor which contributes to this faith is that $\nabla^\mu G_{\mu\nu} \equiv 0$ implies $\nabla^\mu T(m)_{\mu\nu} = 0$, the energy-momentum conservation law. However, this is only necessary but not sufficient for a dynamic solution. Although the 1915 equation gives an excellent description of planetary motion, including the remaining advance of the perihelion of Mercury, this is essentially a test-particle theory, in which the reaction of the test particle is neglected. Thus, the so obtained solutions are not dynamic solutions. As pointed out by Gullstrand [62] such a solution may not be obtainable as a limit of a dynamic solution.

Nevertheless, Einstein, Infeld, and Hoffmann [130] incorrectly assumed the existence of a bounded dynamic solution and deduced the geodesic equation from the 1915 equation. Recently, Feymann [51] made the same incorrect assumption that a physical requirement would be unconditionally applicable. In 1957, Fock [87] pointed out that, in harmonic coordinates, there are divergent logarithmic deviations from expected linearized behavior of the radiation. However, after the discovery that some vacuum solutions are not logarithmic divergent [55], the problems of Einstein's equation was not recognized. Instead, the method of calculation was mistaken as the problem because of the failure to see that the dynamic Einstein equation does not satisfy the principle of causality [65].

To avoid the logarithms, Bondi, et al. [129] and Sachs [131] introduced a new approach to gravitational radiation theory. They used a special type of coordinate system, and they assume the existence of an asymptotic expansion in inverse power of the distance r (from the origin where the isolated source is located in $r \leq a$, which is a positive constant). The approach of Bondi-Sachs was clarified by the geometrical 'conformal' reformulation of Penrose [132]. However, this approach is unsatisfactory, "because it rests on a set of assumptions that have not been shown to be satisfied by a sufficiently general solution of the inhomogeneous Einstein field equation [133]." This approach provides only a definition of a class of space-times that one would like to associate to radiative isolated systems, but neither the global consistency nor the physical appropriateness of this definition has been proven. Moreover, perturbation has given some hints of inconsistency between the Bondi-Sachs-Penrose definition and some approximate solution of the field equation.

There are two other main classes of approach: 1) the post-Newtonian approaches (1/c expansions) and the post-Minkowskian approaches (K expansions). The post-Newtonian approaches are fraught with serious internal consistency problems [133] because they often lead, in higher approximations, to divergent integrals. The post-Minkowskian approach is an extension of the linearization, one may expect that there are some problems related to divergent logarithmic deviations [87]. Moreover, it has unexpectedly been found that perturbative calculations on radiation actually depend on the approach chosen [75].

Mathematically, this non-uniqueness shows, in disagreement with eq. (21), that a dynamic solution of the Einstein equation (6) is unbounded. Also, based on the binary pulsar experiments, it is proven that the Einstein equation does not have any bounded dynamc solution even for weak gravity [7]. This long process is, in part, due to theoretical consistency that was inadequately considered [7, 74, 94, 134]. Moreover, it was not recognized that boundedness of a wave is crucial for its association with a dynamic source. These inadequacies allowed incorrect acceptance of unphysical "time-dependent" solutions as physical waves [65].

Moreover, gravitational radiation is often considered as due to the acceleration in a geodesic alone

[135-137]. It is remarkable that in 1936 Einstein and Rosen [109] are the first to discover this problem of excluding the gravitational wave. However, it was difficult to make an appropriate modification.

The Hulse and Taylor binary pulsar experiments are indispensable for verifying the necessity of the anti-gravity coupling in general relativity [7, 94]. In addition to experimental supports, the Maxwell-Newton Approximation can be derived from physical principles, and the equivalence principle also implies boundedness of a normalized metric in general relativity [8]. A perturbative approach cannot be fully established for the Einstein equation (6) simply because there are no bounded dynamic solutions, which must, owing to radiation, be associated with an anti-gravity coupling.

Nevertheless, Christodoulou and Klainerman [81] claimed to have constructed bounded gravitational (unverified) waves. Obviously, their claim is incompatible with the findings of others. Furthermore, their presumed solutions are incompatible with Einstein's radiation formula and are unrelated to dynamic sources [8, 94]. Thus, they simply have mistaken an unphysical function as a wave [80]. Nevertheless, their book can serve as evidence that the Princeton University can be wrong just as any human institute. Following Einstein's errors, Witten [69] and many of the so-called authorities continue to produce errors in general relativity [143].

Within the theoretical framework of general relativity, however, the gravitational field of a radiating asymptotically Minkowskian system is given by the Maxwell-Newton Approximation [7]. Note that, for the dynamic case, the Maxwell-Newton Approximation is a linearization of the up-dated modified Einstein equation of 1995, but not the Einstein equation, which has no bounded dynamic solution. With the need of rectifying the 1915 Einstein equation established, the exact form of $t(g)_{\mu\nu}$ in the equation of 1995 update [7] is an important problem since a dynamic solution that gives an approximation for the perihelion of Mercury remains unsolved [62]. Moreover, the updated equation shows that the singularity theorems prove only the breaking down of theories of the Wheeler-Hawking school, but not general relativity. This analysis suggests that further confirmation of this approximation is expected.

However, the Wheeler School still has strong influence; and even the MIT open course Phys, 8,033 and Phys, 8.962 are currently filled with their errors.

Due to the influences of the Wheeler School, general relativity is incorrectly believed as effective only for large scale problems. Thus, the study for the applications of general relativity on earth and understanding material structure is neglected or ignored [6, 16]. For example, there are numerous experiments on the weight reduction of a charged capacitor [122, 123]. However, due to the biased view and ignorance of editors of journals such as the Physical Review, and Nature, these experiments are unfairly regarded as due to errors. These experiments are important because they support the charge-mass interaction that is a crucial for the unification of electromagnetism and gravitation [121]. In conclusion, general relativity remains to be completed.

A basic problem is that just as in Maxwell's classical electromagnetism, there is also no radiation reaction force in general relativity. Although an accelerated massive particle would create radiation, the metric elements in the geodesic equation are created by particles other than the test particle. In other words, not only the field equation, but also the equation of motion must be modified. Now, it should be clear that gravitation is not a problem of geometry as the Wheeler School advocated. *Moreover, general relativity is not yet complete, independent of the need of unification due to the existence of the charge-mass interaction.*

The theoretical framework of general relativity is inadequate. In addition to rectification, an extension to include the mass-charge interaction, a repulsive gravitation, is necessary. The mass-charge interaction shows explicitly that the electric energy is not equivalent to mass. Moreover, the five-dimensional relativity proposed by Lo et al. [46] would be a natural possibility.

In current theory, the charge-mass repulsive force would be subjected to electromagnetic screening. Physically, it is unnatural that a neutral force could be screened in such a way. From the viewpoint of the five-dimensional theory, however, the charge-mass repulsive force would be understood as that the charge interacts with a new field created by a mass. Therefore, the repulsive force would not be subjected to such screening. It thus follows that such a force is a perfect test for the existence of a five-dimensional space. Moreover, this can be verified by simply weighing a capacitor before and after charged.

Einstein predicted the weight of a charged capacitor would increase slightly because Einstein believed that the increment of energy would increase the gravitational attraction [127]. However, in a five-dimensional theory, the charge-mass repulsive force is not subjected to screening, and thus would make the charged capacitor lighter. In a charged capacitor, both the positive and the negative charges are concentrated, and thus an effect of the repulsive force would be observed as a lighter weight for the charged capacitor [10].

The charge-mass repulsive force is a fifth force that is independent of the four known forces. Moreover, if the investigation of electric energy leads to a charge-mass repulsive force, the magnetic energy would similarly generate a current-mass force. According to the effect of a magnetic field in general relativity [20, p. 263], it is expected that the current-mass force would be an attractive force. However, details of this issue are beyond the scope of this paper and need further investigation (see Section 13).

The non-equivalence between mass and electromagnetic energy means that the photons must include non-electromagnetic energy (see Section 7). Thus, a very clear experimental verification is desirable. The existence of the charge-mass repulsive force would be a deciding evidence for the non-equivalence between mass and electromagnetic energy. A consequence is that an electron would be slower than a proton falling toward the earth [128].

Gravitation was considered as producing attractive force only, and all the coupling constants were assumed to have the same sign. Recently, it is proven that for the gravitational radiation of binary pulsars the coupling constants must have different signs [7-9]. Now, clearly the electromagnetic energy would not be equivalent to massive energy. Thus, the physical picture provided by Newton is just too simple for a phenomenon as complicated as gravity that relates to everything.

Moreover, misunderstanding $E = mc^2$ actually started from special relativity as Will [28] did. This error leads to the conclusions by Hawking and Penrose that general relativity is inapplicable to microscopic phenomena and inconsistent with quantum theory [2]. Another problem of such a misinterpretation is that the existence of a charge-mass repulsive force was overlooked. Consequently, it was not recognized that general relativity of Einstein was not yet a self-consistent theory. In other words, general relativity is not yet ready for the stage of an overall unification. Thus, it is unrealistic to expect the string theorists to perform a miracle in unification.

Einstein's "covariance principle" was a mistake [37], due to his inadequate ability to distinguish the difference between mathematics and physics. Since the existence of the charge-mass repulsive force is established, the unification of gravitation and electromagnetism is necessary. From the weight reductions of charged capacitors we conclude: 1) $E = mc^2$ is only conditionally valid, and the electromagnetic energy is not equivalent to mass. 2) Einstein's conjecture of unification is established. Moreover, the Einstein equation is invalid for the dynamic case, for which it remains to be rectified and completed in at least two aspects: a) The exact form of the gravitational energy-stress tensor is not known; and b) The radiation reaction force is also not known. Note that, due to the radiation reaction force, considering general relativity as a theory of geometry is invalid.

Moreover, the existence of a dynamic solution requires, as shown in eq. (27), an additional gravitational energy-momentum tensor with an antigravity coupling. Thus, the space-time singularity theorems, which require the same sign for couplings, are actually irrelevant to physics. The positive energy theorem of Schoen and Yau is only for stable solutions because their theorem actually does not include the dynamic solutions. Moreover, $E = mc^2$ is only conditionally valid, and such recognition is crucial to identify the charge-mass interaction. This repulsive force also can potentially explain the Space-Probe Pioneer anomaly [138]. This force would also be useful to detect things underground and/or under water since the strength of such detection can be adjusted with the potential of a charged capacitor [121]. Experimentally, in contrast to the claim of Einstein [127], a piece of heated-up metal has reduced weight [126]. Moreover, since such a force is coupled to the charge square, such a force exists in a five-dimensional theory. It should also be noted that the charge-mass interaction is not included in Quantum mechanics although such an interaction is there.

The experimental confirmation of such an interaction means that Einstein's unification between electromagnet-ism and gravitation is proven valid [126]. Einstein failed to show such unification because of his three shortcomings: 1) He failed to see as Maxwell did

that unification is necessary to create new interactions. 2) He has mistaken that $E = mc^2$ was unconditional. 3) He invalidly rejected repulsive gravitation. Hence, it turns out that Einstein is the biggest winner from the rectification of his theories.[28] It is up to us to continue the remarkable work that Einstein started.

Acknowledgments

This paper is dedicated to Prof. P. Morrison of MIT for unfailing guidance for over 15 years. The author gratefully acknowledges stimulating discussions with Prof. A. J. Coleman, Prof. I. Halperin, and Prof. J. E. Hogarth. The author wishes to express his appreciation to S. Holcombe for valuable suggestions. This publication is supported by Innotec Design, Inc., U.S.A.

Appendix: On Interpretations of Hubble's Law and Einstein's Theory of Measurement

Hubble's law is often considered as the observational evidence of an expanding universe. However, Hubble's Law need not be related to the Doppler redshifts of the light from receding Galaxies. For instance, such a receding velocity is incompatible with the local light speeds used in deriving the light bending. Thus, the notion of an expanding universe is based on a questionable assumption that a local distance is similar to that of a mathematical Riemannian space. However, such an assumption has been pointed out by Whitehead [40] and also proven as theoretically invalid. Thus, the notion of an expanding universe could be just a mathematical illusion.

Currently, Hubble's law is often considered as a manifestation of the Doppler red shift of the light from the receding Galaxies [2]. Then, the further a galaxies is from the Milky Way, the faster it appears to receding. However, Hubble himself rejected this interpretation and concluded in 1936 that the Galaxies are actually stationary [139].

It will be shown that Hubble's Law need not be related to the Doppler redshifts of the light from receding Galaxies. It is pointed out that an implicit assumption, which leads to no expansion for the space coordinates, must be used. Moreover, the receding velocity is incompatible with the light speeds used in deriving the light bending. In short, the notion of expanding universe is a production due to an inadequate understanding of a physical space. Thus, it is questionable that such a universe is related to the reality.

A.1. *Hubble's Law*

Hubble observed from light emitted by galaxies that the redshifts S are linearly proportion to the distance L from the Milky Way as,

$$S = H L \tag{A1}$$

where H is the Hubble constant although the redshifts of distant galaxies will deviate from this linear law slightly.

A.2. *The Redshifts*

In terms of a theory from general relativity, it is well known that this law can be derived with the following metric [29],

$$ds^2 = -d\tau^2 + a^2(\tau)\{dx^2 + dy^2 + dz^2\}, \tag{A2}$$

since

$$S = \frac{\lambda_2 - \lambda_1}{\lambda_1} = \frac{\omega_1}{\omega_2} - 1 = \frac{a(\tau_2)}{a(\tau_1)} - 1, \tag{A3}$$

where ω_1 is the frequency of a photon emitted at event P_1 at time τ_1, and ω_2 is the frequency of the photon observed at P_2 at time τ_2 [2]. Moreover, for nearby galaxies, one has

$$a(\tau_2) \approx (\tau_1) + (\tau_2 - \tau_1)\tilde{a}. \quad \tau_2 - \tau_1 \approx R \tag{A4}$$

Thus,

$$S = \frac{\tilde{a}}{a}L = H L, \quad \text{and} \quad H = \frac{\tilde{a}}{a}. \tag{A5}$$

Note that Hubble's Law need not be related to the Doppler redshifts. (Hubble rejected such an interpretation [139].)

A.3. *Hubble's Law and the Doppler Redshifts*

If one chooses to define the distance between two points as

$$R = \int_1^2 a(\tau)\sqrt{dx^2 + dy^2 + dz^2} = a(\tau)\,L,$$

where $\quad L = \int_1^2 \sqrt{dx^2 + dy^2 + dz^2} \tag{A6}$

Then

$$v = \frac{dR}{d\tau} = \frac{da}{d\tau}L + \frac{dL}{d\tau}a = \frac{da}{d\tau}\frac{R}{a} = HR, \tag{A7}$$

Thus,

$$v = S. \tag{A8}$$

This means that the redshifts could be superficially considered as a Doppler effect.

A.4. *Remarks*

However, if we define the distance as L, there is actually no receding velocity since L is fixed ($dL/d\tau = $

0). Thus, whether Hubble's Law represents the effects of an expanding universe is a matter of the interpretation of the local distance. From the above analysis, the crucial point is what a valid physical velocity in a physical space is.

It should be noted that $dL/d\tau = 0$ means that the space coordinates are independent of physics. In other words, the physical space has a Euclidean-like structure [37], which is independent of time. However, since L between any two space-points is fixed, the notion of an expanding universe, if it means anything, is just an illusion. Moreover, the validity of (A6) as the physical distance has no known experimental supports since it is not even clearly measurable. Moreover, a problem is that the notion of velocity in (A7) would be incompatible with the light speeds in the calculation of light bending experiment.

A.5. *The Coordinates of an Einstein Physical Space, and Definition of Velocity*

In mathematics, the Riemannian space is often embedded in a higher dimensional flat space [140]. Then the coordinates dx^μ are determined by the metric through the metric,

$$ds^2 = g_{\mu\nu}\,dx^\mu dx^\nu, \text{ or } -g_{tt}dt^2 + g_{ij}dx^i dx^j \tag{A9}$$

such as the surface of a sphere in a three dimensional Euclidean space. For a physical space, however, there are insufficient conditions to do so. Since the metric is a variable function, it is impossible to determine the coordinates with the metric. In view of this, the coordinates must be physically independent of the metric. Moreover, it has been proven from the theoretical framework of general relativity [37] that a frame of reference with the Euclidean-like structure must exist for a physical space.

For a spherical mass distribution with the center at the origin, the metric with the isotropic gauge is,

$$ds^2 = -[(1 - M\kappa/2r)^2/(1 + M\kappa/2r)^2]\,c^2dt^2 + (1 + M\kappa/2r)^4 (dx^2 + dy^2 + dz^2) \tag{A10}$$

where $\kappa = G/c^2$ ($G = 6.67 \times 10^{-8}$ erg cm/gm²), M is the total mass, and $r = \sqrt{x^2 + y^2 + z^2}$. Then, if the equivalence principle is satisfied, the light speeds are determined by $ds^2 = 0$ [4, 5], i.e.,

$$\frac{\sqrt{dx^2 + dy^2 + dz^2}}{dt} = c\,\frac{1 - M\kappa/2r}{(1 + M\kappa/2r)^3} \tag{A11}$$

However, such a definition of light speeds is incompatible with the definition of velocity (A7). *Since this light speed is supported by observations, definition (A7) is invalid in physics.*

Nevertheless, Liu [141] has defined light speeds, which is more compatible with (A7), as

$$\frac{\sqrt{g_{ij}dx^i dx^j}}{dt} = c\,\frac{1 - M\kappa/2r}{1 + M\kappa/2r} \qquad (A12)$$

for metric (A10). However, (A12) implies only half of the deflection implied by (A11) [4, 5].

The above analysis also explains why many current theorists insist on that the light speeds are not defined even though Einstein defined them clearly in his 1916 paper as well as in his book, "The Meaning of Relativity". They might argued that the light speeds are not well defined since diffeomorphic metrics give different sets of light speeds for the same frame of reference. However, they should note that Einstein defines light speeds after the assumption that his equivalence principle is satisfied [4, 5]. Different metric for the same frame of reference means only that at most only one of such metrics is physically valid.

Moreover, it has been proven that the Maxwell-Newton Approximation gives the valid first order approximation of the physical metric, the first order of the physically valid light speeds are solved [37]. Since metric (A10) is compatible with the Maxwell-Newton approximation, the first order of light speed (A11) is valid in physics.

Thus, the groundless speculation that local light speeds are not well defined is proven incorrect. In essence, the velocity definition (A7), which leads to the notion of the Doppler redshifts, has been rejected by experiments. Nevertheless, some skeptics might prefer to accept formula (A6) after light speed (A11) is confirmed by the experiment of local light speeds [37].

A. 6. *Discussions*

One may ask what causes such redshifts that are roughly proportional to the distances from the observer. One possibility is that the scatterings of a light ray along its path to the observer. In physics, it is known that different scatterings are common causes for losing energy of a particle, and for the case of photons it means redshifts. Unfortunately, to test such a conjecture is not possible because no current theory of gravity is capable of handling the inelastic scatterings of lights.

Nevertheless, the assumption that observed redshifts could be due to inelastic scatterings may help to explain some puzzles of observed facts [142]. For instance, younger objects such as star forming galaxies have higher intrinsic redshifts, and objects with the same path length to the observer have much different redshifts while all parts of the object have about the same amount of redshifts. For those interested in alternative cosmology theories, there are the plasma universe model [143] and other models [144].

Endnotes:

1) The editors of General Relativity and Gravitation considers the claims of the Wheeler School as "well-established science", but were unable to provide supporting evidence [March 8, 2012]. Note that since there is no bounded dynamic solution for the Einstein equation [67], the thesis of A. Ashtekar (editor-in-chief), *"Asymptotic Structure of the Gravitational Field at Spatial Infinity"*, just inherits the errors of Wald [76]. Moreover, he failed to see that the photons must include gravitational energy [107]. C. M. Will, editor-in-chief of Classical and Quantum Gravity, continues to ignore the errors of the Wheeler School [145].

2) Although a Nobel Prize has been awarded to related scientific claims, this is not yet a confirmation in sciences (see Appendix).

3) The Wheeler School also failed to respond to the challenge of Bondi, Pirani, & Robinson [55]; and were unable to rectify their error on local time shown in their eq. (40. 14); and made invalid claims on dynamic solutions and physical principles [145].

4) This experiment [23] is ignored by the Wheeler School or they simply were unaware of this.

5) It is surprising that "expert" Thorne [20] also made such a factual error. This shows clearly that the Wheeler School is unreliable.

6) Nevertheless, the 1993 Nobel Committee was unaware of that Einstein's equivalence principle has been verified [64].

7) Like other theoretical physicists, Pauli [24] and Misner et al. [1] also did not have adequate training in pure mathematics [145].

8) The misinterpretation of Misner et al. [1] creates the so-called Lorentz invariance, being tested by Chung, Chiow, Herrmann, Chu, & Müller [146].

9) Being a student of Oppenheimer, P. Morrison of MIT has a very sharp ability in distinguishing the physics from mathematics.

10) Yang-Mills-Shaw [42, 43].made a crude proposal, but the underlying idea of total gauge invariance is invalid [45].

11) The Wheeler School [145] failed to defend the requirement for weak gravity to meet the challenge of Bondi et al. [55].

12) While being a good applied mathematician, 't Hooft has a poor understanding in physics as shown in his 1999 Nobel Lecture [61], 't Hooft [59] claimed that many of his colleagues agree with him, but this only means they make the same error.

13) Such an inconsistency has been discovered, and Einstein's derivation was not repeated in most textbooks [145].

14) A main error of Einstein, Infeld, & Hoffmann [100], Damour [97], Misner et al, [1], Wald [2], Will [28] and etc. is that they are unaware of that the mathematical existence of a bounded dynamic solution needs to be proved [145].

15) Bertschinger [63] did not know that, for the dynamic case, the linearized equation and the non-linear Einstein are actually independent equations since the non-linear equation has no bounded dynamic solutions [71].

16) The unique sign of couplings [2] was accepted because the formula $E = mc^2$ was believed to be unconditional.

17) Members of the selection committee seem to be very careless. Had the Selection Committee tried to find an example of the dynamic solution that could support the claims of Christodoulou, they would have found his errors [82].

18) Christodoulou & Klainerman [81] were unaware that their set of solutions may have only static physical solutions [80]. Obviously, Christodoulou was still not aware of this when he received his half Shaw Prize in 2011 [82].

19) Nobel Laureate 't Hooft [90] also believe that linearization is unconditionally valid as Bertschinger [63] did. However, the error is probably originated from the book of Christodoulou & Klainerman [81].

20) For a thorough discussion on the relation between the mass and the total energy of a particle, one can read the paper of L. B. Okun [147]. However, Okun did not understand that the electromagnetic energy is not equivalent to mass [148].

21) Ludwig D. Faddeev, the Chairman of the Fields Medal Committee, wrote ("On the work of Edward Witten"):

"Now I turn to another beautiful result of Witten – proof of positivity of energy in Einstein's theory of gravitation. Hamiltonian approach to this theory proposed by Dirac in the beginning of the fifties and developed further by many people has led to the natural definition of energy. In this approach a metric γ and external curvature h on a space-like initial surface $S^{(3)}$ embedded in space-time $M^{(4)}$ are used as parameters in the corresponding phase space. These data are not independent. They satisfy Gauss-Codazzi constraints – highly non-linear PDE, The energy H in the asymptotically flat case is given as an integral of indefinite quadratic form of $\nabla\gamma$ and h. Thus, it is not manifestly positive. The important statement that it is nevertheless positive may be proved only by taking into the account the constraints – a formidable problem solved by Yau and Schoen in the late seventy as Atiyah mentions, 'leading in part to Yau's Fields Medal at the Warsaw Congress'.

Witten proposed an alternative expression for energy in terms of solutions of a linear PDE with the coefficients expressed through γ and h ....."

22) Michael Francis Atiyah has been leader of the Royal Society (1990-1995), master of Trinity College, Cambridge (1990-1997), chancellor of the University of Leicester (1995-2005), and President of the Royal Society of Edinburgh (2005-2008). Since 1997, he has been an honorary professor at the University of Edinburgh (Wikipedia). Apparently, Atiyah does not understand the physics and the non-existence of a dynamic solution for the Einstein equation [70].

23) This is a case that the static Einstein equation can predict beyond the Maxwell-Newton Approximation [121]. However, this metric was subjected to severe misinterpretations because theorists, including Einstein, did not accept a repulsive gravitation.

24) The Cylindrical Condition is defined as that for all physical quantities are functions of four variables [120].

25) At MIT, only P. Morrison understood the non-existence of bounded dynamic solution. He also went to Princeton to question J. A. Taylor on the justification of their calculation on radiation. As expected, Taylor failed to provide a valid justification [64].

26) The 1911 assumption is well-known to be incorrect after the 1919 British expeditions [4].

Thus, there is no rational reason to take the 1911 assumption of equivalence between acceleration and Newtonian gravity as the reference for Einstein's equivalence principle, instead of his statements in his 1916 paper and his book. Such acts support the suspicion that the Wheeler School had planned to get rid of Einstein's equivalence principle because of its conflict with the covariance principle.

27) I have reported these to MIT President Hockfield and the subsequent President Reif. However, their promises of up-grading the gravitational education were not supported by the Physics Department that was dominated by the Wheeler School.

28) Because Einstein was unable to recognize the limitations and errors of his earlier work, he failed to make progress in relativity after he arrived in the US. Understandably, Einstein refused to extend his life by available medicine [149] by claiming "It is tasteless to prolong life artificially, I have done my share, it is time to go. I will do it elegantly." Had Einstein known that he was very close to his unification, would he still be that willing to go?

References:

1. C. W. Misner, K. S. Thorne, & J. A. Wheeler, *Gravitation* (W. H. Freeman, San Francisco, 1973).
2. R. M. Wald, *General Relativity* (The Univ. of Chicago Press, Chicago, 1984).
3. On Achievements, Shortcomings and Errors of Einstein, *International Journal of Theoretical and Mathe-matical Physics*, Vol.4, No.2, 29-44 (2014).
4. A. Einstein, H. A. Lorentz, H. Minkowski, H. Weyl, *The Principle of Relativity* (Dover, New York, 1923).
5. A. Einstein, *The Meaning of Relativity* (Princeton Univ. Press, 1954).
6. C. Y. Lo, On Gauge Invariance in Physics & Einstein's Covariance Principle, Phys. Essays, **23** (3), 491-499 (Sept. 2010).
7. C. Y. Lo, *"Einstein's Radiation Formula and Modifications to the Einstein Equation,"* Astrophysical Journal **455**, 421-428 (Dec. 20, 1995); Editor S. Chandrasekhar suggests the appendix therein.
8. C. Y. Lo, *"Compatibility with Einstein's Notion of Weak Gravity: Einstein's Equivalence Principle and the Absence of Dynamic Solutions for the 1915 Einstein Equation,"* Phys. Essays **12** (3), 508-526 (1999).
9. C. Y. Lo, *"On Incompatibility of Gravitational Radiation with the 1915 Einstein Equation,"* Phys. Essays **13** (4), 527-539 (December, 2000).
10. C. Y. Lo, *"The Invalid Speculation of m = E/c², the Reissner-Nordstrom Metric, and Einstein's Unification,"* Phys. Essays, **25** (1), 49-56 (2012).
11. C. Y. Lo, Comments on Misunderstandings of Relativity, and the Theoretical Interpretation of the Kreuzer Experiment, Astrophys. J. **477,** 700-704 (1997).
12. C. Y. Lo, The Necessity of Unifying Gravitation and Electromagnetism, Mass-Charge Repulsive Effects, and the Five Dimensional Theory, Bulletin of Pure and Applied Sciences, **26D** (1), 29 - 42 (2007a).
13. A. Einstein & W. Pauli, Ann. Math. **44**, 133 (1943).
14. V. A. Fock, *The Theory of Space Time and Gravitation* (Pergamon Press, 1964). The Russian edition was published in 1955 as part of the mud throwing campaign to discredit Einstein, after his death.
15. The 1993 Press Release of the Nobel Prize Committee (The Royal Swedish Academy of Sciences, Stockholm, 1993).
16. C. Y. Lo, Bulletin of Pure and Applied Sciences, **26D** (2): 73-88 (2007).
17. R. C. Tolman, *Relativity, Thermodynamics, and Cosmology* (Dover, New York 1987).
18. J. P. Hsu, & L. Hsu, *Chinese J. of Phys.*, Vol. 35 (No. 4): 407-417 (1997).
19. P. G. Bergmann, *Introduction to the Theory of Relativity* (Dover, New York, 1976), pp. 156, 159.
20. K. S. Thorne, *Black Holes and Time Warps* (Norton, New York, 1994), p. 105.
21. C. Y. Lo, Phys. Essays **18** (4), 547-560 (December, 2005).
22. L. D. Landau & E. M. Lifshitz, *Classical Theory of Fields* (Addison-Wesley, Reading Mass, 1962).
23. W. Kundig, Phys. Rev, 129, 2371 (1963).
24. W. Pauli, *Theory of Relativity* (Pergamon Press, London, 1971).
25. J. Norton, "What was Einstein's Principle of Equivalence?" in Einstein's Studies Vol.1: *Einstein and the History of General Relativity,* Eds. D. Howard & J. Stachel (Birkhäuser, Boston, 1989).
26. A. Einstein, On the influence of Gravitation on the propagation of light, Annalen der Physik, 35, 898-908 (1911).
27. J. L. Synge, *Relativity: The General Theory* (North-Holland, Amsterdam, 1971), pp. IX–X.
28. C. M. Will, *Theory and Experiment in Gravitational Physics* (Cambridge University, Cambridge. 1981).
29. H. C. Ohanian & R. Ruffini, *Gravitation and Spacetime* (Norton, New York, 1994).

30. C. Y. Lo, Phys. Essays, **23** (2), 258-267 (2010); C. Y . Lo, Phys. Essays **20** (3), 494–502 (Sept. 2007).

31. C. Y. Lo, Phys. Essays **16** (1), 84-100 (March 2003).

32. C. Y. Lo, Bulletin of Pure and Applied Sciences, **26D** (2): 73-88 (2007).

33. Zhou, Pei-Yuan, in *Proc. of the Third Marcel Grossmann Meetings on Gen. Relativ.* ed. Hu Ning, Sci. Press/North Holland. (1983), 1-20.

34. P. Y. Zhou, Proc. of the Internat. Symposium on Experimental Gravitational Physics, Guang Zhou, China (1987).

35. C. Y. Lo, Bulletin of Pure and Applied Sciences, **27D** (1), 1-15 (2008).

36. C. Y. Lo, **16th Annual Natural Philosophy Alliance Conference**, Univ. of Connecticut, Storrs, May 25-29, 2009.

37. C. Y. Lo, Chinese J. of Phys. (Taipei), **41** (4), 233-343 (August 2003).

38. A. S. Eddington, *The Mathematical Theory of Relativity* (Chelsea, New York, 1975), p. 10.

39. C. Y. Lo, Phys. Essays, **18** (1), 112-124 (2005).

40. A. N. Whitehead, *The Principle of Relativity* (Cambridge Univ. Press, Cambridge, 1962).

41. C. Y. Lo, On Interpretations of Hubble's Law and the Bending of Light, Progress in Phys., Vol. 1, 10-13 (Jan., 2006).

42. C. N. Yang & R. L. Mills, Phys. Rev. **96**, 191 (1954).

43. Ron Shaw, "The Problem of Particle Types and Other Contributions to the Theory of Elementary Particles," Ph. D. thesis, Cambridge University (1955).

44. Y. Aharonov & D. Bohm, Phys. Rev. **115**, 485 (1959).

45. S. Weinberg, *The Quantum Theory of Fields* (Cambridge University Press, Cambridge, 2000).

46. C. Y. Lo, G. R. Goldstein, & A. Napier, Hadronic J. **12**, 75 (1989).

47. C. Y. Lo, Phys. Essays, **5** (1), 10-18 (1992).

48. J. Bodenner & C. M. Will, Am. J. Phys. 71 (8), 770 (August 2003).

49. J. M.Gérard & S.Piereaux, The Observable Light Deflection Angle, arXiv:gr-qc/9907034 v18 Jul (1999).

50. C. Y. Lo, Bulletin of Pure and Applied Sciences, **28D** (1), 67-85 (2009).

51. R. P. Feynman, *The Feynman Lectures on Gravitation* (Addison-Wesley, New York, 1995).

52. C. Y. Lo, Proc. IX International Sci. Conf. on **'Space, Time, Gravitation,'** Saint-Petersburg, August 7-11, 2006.

53. C. Y. Lo, Linearization of the Einstein Equation and the 1993 Press Release of the Nobel Prize in Physics, in Proc. of 18 th Annual Natural Philosophy Alliance Conf., Vol. **8**, 354-362, Univ. of Maryland, USA. 6-9 July (2011).

54. C. Y. Lo, **Math., Physics and Philo. in the Interpretations of Relativity Theory II,** Budapest, 4-6 Sept. 2009.

55. H. Bondi, F. A. E. Pirani, & I. Robinson, Proc. R. Soc. London A **251**, 519-533 (1959).

56. R. Penrose, Rev. Mod. Phys. 37 (1), 215-220 (1965).

57. C. Y. Lo, International Meeting on Physical Interpretation of Relativity Theory, Imperial College, London, Sept. 12-15, 2008b; Bull. of Pure and App. Sci., **29D** (2), 81-104 (2010).

58. C. Y. Lo, On Physical Invalidity of the "Cylindrical Symmetric Waves" of 't Hooft, Phys. Essays, **24** (1), 20-27 (2011).

59. G. 't Hooft, **"Strange Misconceptions of General Relativity"**, (there are other problems beyond those addressed in reference [63] therein) http://www.phys.uu.nl/~thooft/gravitating_misconceptions.html (2011).

60. C. Y. Lo, The Principle of Causality and Einstein's Requirement for Weak Gravity, versus Einstein's Covariance Principle, Bulletin of Pure and Applied Sciences, **29D** (2), 81-104 (2010).

61. G. 't Hooft, "A Confrontation with Infinity", Nobel Lecture, December, 1999.

62. A. Gullstrand, Ark. Mat. Astr. Fys. 16, No. 8 (1921); ibid, Ark. Mat. Astr. Fys. 17, No. 3 (1922).

63. Edmund Bertschinger, *Cosmological Dynamics,* Department of Physics MIT, Cambridge, MA 02139, USA.

64. C. Y. Lo, On the Nobel Prize in Physics, Controversies and Influences, GJSFR vol. 13-A Issue 3 Ver. 1.0, 59-73 (June 2013).

65. C. Y. Lo, The Gravitational "Plane Waves" of Liu and Zhou and the Nonexistence of Dynamic Solutions for Einstein's Equation, Astrophys. Space Sci., **306**: 205-215 (2006).

66. S. Hod, A simplified two-body problem in general relativity, IJMPD Vol. 22, No. 12 (2013).

67. C. Y. Lo, On the Question of a Dynamic Solution in General Relativity, J. of Space Exploration **4** (2013).

68. R.Schoen and S.-T.Yau, Proof of the Positive Mass Theorem. II, Commun. Math. Phys. **79**, 231 (1981).

69. E. Witten, "A New Proof of the Positive Energy Theorem," Commun. Math. Phys., **80**, 381 (1981).

70. C. Y. Lo, The Errors in the Fields Medals, 1982 to S. T. Yau, and 1990 to E. Witten, GJSFR vol. 13-F, Iss. 11, Ver. 1.0 (2014).

71. C. Y. Lo, The Non-linear Einstein Equation and Conditionally Validity of its Linearization, Intern. J. of Theo. and Math. Phys., Vol. 3, No.6 (2013).

72. Pring Pring F, The Royal Society, "Board Member's Comments" (Jan. 8, 2007).

73. H. Y. Liu, & P.-Y. Zhou, Scientia Sincia (Series A) 1985, XXVIII (6) 628-637.

74. C. Y. Lo, Duality of Electromagnetic Waves, Causality on Gravity, and the Necessity of Modification in the Einstein Equation, Phys. Essays **10** (3), 424-436 (September, 1997).

75. N. Hu, D.-H. Zhang, & H.-G. Ding, Acta Phys. Sinica, 30 (8), 1003-1010 (Aug. 1981).

76. C. Y. Lo, Some Mathematical and Physical Errors of Wald on General Relativity, GJSFR vol. 13-A Iss. 2 Ver. 1.0, (April 2013).

77. S. Weinberg, *Gravitation and Cosmology* (John Wiley, New York, 1972), p. 273.

78. A. Einstein, Sitzungsberi, Preuss, Acad. Wis. 1918, 1: 154 (1918).

79. K. S. Thorne, *Black Holes and Time Warps* (Norton, New York, 1994), p. 105.

80. C. Y.Lo, Phys.Essays 13(1),109-120 (March 2000)

81. D. Christodoulou & S. Klainerman, *The Global Nonlinear Stability of the Minkowski Space* (Princeton. Univ. Press, 1993).

82. C. Y. Lo, Comments on the 2011 Shaw Prize in Mathematical Sciences, -- an analysis of collectively formed errors in physics, GJSFR Vol. 12-A Issue 4 (Ver. 1.0) (June 2012).

83. C. Y. Lo, Einstein's Radiation formula and Modifications in General Relativity, The Second William Fairbank Conference, Hong Kong Polytechnic, Hong Kong Dec. 13-16 (1993).

84. Volker Perlick, Zentralbl. f. Math. (827) (1996) 323, entry Nr. 53055.

85. Volker Perlick (republished with an editorial note), Gen. Relat. Grav. 32 (2000).

86. C. Y. Lo, Phys. Essays, 23 (2), 258-267 (2010).

87. V. A. Fock, Rev. Mod. Phys. 29, 325 (1957).

88. D. Kramer, H. Stephani, E. Herlt, & M. MacCallum, *Exact Solutions of Einstein's Field Equations*, ed. E. Schmutzer (Cambridge Univ. Press, Cambridge, 1980).

89. W.Kinnersley, "Recent Progres in Exact Solutions" in General Relativity and Gravitation (Proceedings of GR7, Tel-Aviv 1974) ed. G. Shaviv, and J.Rosen (Wiley New York, London, 1975).

90. C. Y. Lo, On Physical Invalidity of the "Cylindrical Symmetric Waves" of 't Hooft, Phys. Essays, **24** (1), 20-27 (2011).

91. C. Y. Lo, The Principle of Causality and the Cylindrically Symmetric Metrics of Einstein & Rosen, Bulletin of Pure and Applied Sciences, **27D** (2), 149-170 (2008).

92. H. W. Ellis and R. R. D. Kempt remarks in their 1964 lectures at Queen's University.

93. J. E. Hogarth, 1953 Ph. D. Thesis, Dept. of Math., Royal Holloway College, Univ. of London, p. 6.

94. C. Y. Lo, The Question of Theoretical Self-Consistency in General Relativity: on Light Bending, Duality, the Photonic Energy-Stress Tensor, and Unified Polarization of the Plane-Wave Forms, Phys. Essays **12** (2), 226-241 (June, 1999).

95. R. A. Hulse & J. H. Taylor, Astrophys. J. Lett. **65,** L51 (1975).

96. C. Y.Lo, in Proc. Sixth Marcel Grossmann Meeting On General Relativity, 1991, ed. H. Sato & T. Nakamura, 1496 (World Sci., Singapore, 1992).

97. T. Damour, "The Problem of Motion in Newtonian and Einsteinian Gravity" in *300 Years of Gravitation* edited by S. W. Hawking and W. Israel (Cambridge Univ. Press., Cambridge, 1987).

98. T. Damour & B. Schmidt, J. Math. Phys. **31** (10), 2441-2453 (October, 1990).

99. P. T. Chruscie, M. A. H. McCallum, & D. B. Singleton, Phil. R. Soc. Lond. A **350,** 113 (1995).

100.A. Einstein, L. Infeld, and B. Hoffmann, Annals of Math. 39 (1), 65-100 (Jan. 1938).

101.H. Bondi, M. G. J. van der Burg, and A. W. K. Metzner, Proc. R. Soc. Lond. A **269,** 21 (1962).

102.T. Damour & J. H. Taylor, Astrophys. J. **366**: 501-511 (1991).

103.T. Damour & J. H. Taylor, Phys. Rev. D, **45** (6), 1840-1868 (1992).

104.H. A. Lorentz, Versl gewone Vergad Akad. Amst., vol. 25, p. 468 and p. 1380 (1916).

105.T. Levi-Civita, R. C. Accad Lincei (5), vol. 26, p. 381 (1917).

106.A. Einstein, Sitzungsber Preuss. Akad. Wiss, vol. 1, p. 167 (1916).

107.C. Y. Lo, *"Completing Einstein's Proof of E = mc²,"* Progress in Phys., Vol. 4, 14-18 (2006).

108.C. Y. Lo, The Gravity of Photons and the Necessary Rectification of Einstein Equation, Prog. in Phys., V. 1, 46-51 (2006).

109.A. Einstein & N. Rosen, J. Franklin Inst. **223**, 43 (1937).

110.A. Pais, *'Subtle is the Lord..'* (Oxford Univ. Press, New York, 1996).

111.*Einstein's Miraculous Year*, edited by John Stachel (Princeton University Press, Princeton 1998).

112.C.Y. Lo, Chin. Phys., **16** (3) 635-639 (March 2007).

113.C. Y. Lo, Bulletin of Pure and Applied Sciences, Vol. 25D, No.1, 41-47, 2006.

114.C. L. Pekeris, Proc. Nat. L Acad. Sci. USA Vol. 79, pp. 6404-6408 (1982).

115.L. Herrera, N. O. Santos and J. E. F. Skea, Gen. Rel. Grav. Vol. 35, No. 11, 2057 (2003).

116. E. T. Whittaker, Proc. R. Soc. (London) A **149**, 384, (1935).

117. R. Tolman, Phys. Rev. **35**, 875 (1930).

118. C. Y. Lo & C. Wong, Bull. of Pure and Applied Sciences, Vol. 25D (No.2), 109-117 (2006).

119. D. Yu. Tsipenyuk, V. A. Andreev, Physical Interpretations of the Theory of Relativity Conference (Bauman Moscow State Technical University, Moscow 2005).

120. Th. Kaluza Sitzungsber, Preuss. Akad. Wiss. Phys. Math. Klasse 966 (1921).

121. C. Y. Lo, Gravitation, Physics, and Technology, Physics Essays, **25** (4), 553-560 (Dec. 2012).

122. Takaaki Musha, "Explanation of the Dynamical Biefeld-Brown Effect from the Standpoint of the ZPF Field, JBIS, vol. 61, PP 379-384, 2008.

123. T. Valone, *Electro Gravitics II* (Integrity Research Institute, Washington DC, 2008).

124. W. Q. Liu, private communication (August 2007).

125. V. Tarko, The First Test that Proves General Theory of Relativity Wrong (http://news.softpedia.com/news/The-First-Test-That-Proves-General-Theory-of-Relativity-Wrong-20259.shtml, 2006) (Accessed March 24, 2006).

126. C. Y. Lo, On the Weight Reduction of Metals due to Temperature Increments, GJSFR Vol. 12 Issue 7 (Ver. 1.0) (Sept. 2012).

127. A. Einstein, 'E = MC^2' (1946), *Ideas and Opinions* (Crown, New York, 1982).

128. C. Y. Lo, "Could Galileo Be Wrong?", Phys. Essays, **24** (4), 477-482 (2011).

129. H. Bondi, M. G. J. van der Burg, and A. W. K. Metzner, Proc. R. Soc. Lond. A **269**, 21 (1962).

130. A. Einstein, L. Infeld, and B. Hoffmann, Annals of Math. **39** (1), 65-100 (Jan. 1938).

131. R. K Sachs, Proc. R. Soc. Lond. A **270**, 103 (1962).

132. R. Penrose, Proc. R. Soc. Lond. A **284**, 159 (1965).

133. L. Blanchet and T. Damour, Phil. Trans. R. Soc. Lond. A, **320**, 379-430 (1986).

134. C. Y.Lo, Phys. Essays, 11(2), 264-272 (June 1998).

135. J. N. Goldberg, Phys. Rev. **89**, 263 (1953).

136. A. E. Scheidegger, Phys. Rev. **99**, 1883 (1955).

137. L. Infeld., Rev. Mod. Phys. **29**, 398 (1957).

138. C. Y. Lo, The Unification of Gravitation and Electromagnetism and the Cause of the Pioneer Anomaly Discovered by NASA, J. of Space Exploration **5** (2014).

139. G. J. Whitrow, *"Edwin Powell Hubble," Dictionary of Scientific Biography*, New York, Charles Scribner's Sons, Vol 5, 1972, p. 532

140. P. A. M. Dirac, *General Theory of Relativity* (John Wiley, New York, 1975).

141. Liu Liao, *General Relativity* (High Education Press, Shanghai, 1987), pp 26-30.

142. Halton Arp, Progress in Physics, vol. 3, 3-6 (October, 2005).

143. E. J. Lerner, *The Big Bang Never Happened* (Vintage, New York 1992).

144. H. Kragh*, Cosmology and Controversy* (Princeton Univ. Press, 1999).

145. C. Y. Lo, Errors of the Wheeler School, the Distortions to General Relativity and the Damage to Education in MIT Open Courses in Physics, GJSFR Vol. 13 Issue 7 Version 1.0 (2013).

146. K.-Y. Chung, S.-w. Chiow, S. Herrmann, Steven Chu, and H. Müller, "*Atom interferometry tests of local Lorentz invariance in gravity and electrodynamics,*" Phys. Rev. D **80**, 016002 (July 2009).

147. L. B. Okun, The concept of mass (mass, energy, relativity),Usp. Fiz. Nauk 158, 511-530 (July 1989)

148. L. B.Okun, The Einstein formula: E_0 = mc^2.Isn't the Lord laughing?, Uspekhi 51 (5), 513-527 (2008).

149. Walter Isaacson, *Einstein- His Life and Universe-*(Simon &Schuster, New York, 2008), p. 542.

VARIABLE SPEED OF LIGHT IN 3-DIMENSIONAL EUCLIDEAN SPACE

NINA SOTINA
Ph.D. in physics
e-mail: nsotina@gmail.com

NADIA LVOV
Essex County College, NJ
e-mail: lvov@essex.edu

The speed of light according to special relativity has the same value c with respect to a distant star, with respect to the Earth or with respect to a moving source. Special relativity explains this paradox through kinematics by proposing that our space is 4-dimensional pseudo-Euclidean and, hence, that the classical law of velocity addition is not correct. In this work we showed that the constancy of the speed of light observed in experiments can be explained remaining in the framework of the model of the three-dimensional Euclidean space and the classical law of velocity addition. But in this case we have to accept the existence of some 'hidden' dynamics that leads to a change in the velocity of light (photon) within the same frames of reference (for example, leaving a source with a velocity $c + u$, where u is a velocity of the source with respect to the Earth, the speed of photon changes its value to the value c near the Earth's surface as is observed in experiments). We show mathematically that the transverse Doppler Effect can be used in support of such hypothesis (note, that the transverse Doppler Effect is still considered the main arguments in favor of relativity kinematics). Another observation that supports the hypothesis that the speed of light changes within a physical frame of reference is astronomical observations of binary stars.

1. Light still remains a "dark" issue in physics.

The speed of light according to **special relativity** (SR) has the same value c with respect to any inertial frame of reference. Wherein the inertial frames of reference are understood to be such frames of references that are related through Lorenz transformation. An attempt to build an alternative physical model in 3-dimensional Euclidean space brings us back to a classical problem: in what frame of reference does light travel with the speed c? More than 100 years passed since the time this problem brought the so-called "crisis in physics" that was settled with the development of SR. During this time new ideas emerged and new experiments were performed among which there were some "problematic" experiments that contradicted the SR. However, it is conventional in the scientific community to consider a phenomenon as established provided it has been confirmed by several well-known independent laboratories. For various reasons the "problematic" experiments were not repeated in these laboratories.

From the model of the three-dimensional Euclidean space and independent time it follows that the speed of light in various geometric frames of reference may have any value. However, this conclusion requires a more detailed discussion when the real physical frames of reference are considered wherein experiments are conducted, specifically, those that demonstrate the

invariance of the speed of light. For example, if we suppose that propagation of light is some process in a medium (aether) then the motion of the aether itself with respect to a given frame of reference should be taken into consideration too.

It has been established that light transfers energy from one physical body, the source, to another, the receiver, in discrete increments, that is, quanta. However, among physicists there is no unified point of view for the description of the material carrier of the quantum, that is, the photon. There are three types of photon that are usually used in descriptions of the optical experiments demonstrating quantum properties of light [1]. The difference in usage of the term "photon" reflects the difference in interpretation of the results of such experiments.

1) ***The C-photon*** is a classical wave packet, that is, spatially localized, quasi monochromatic electromagnetic radiation carrying a quantum of energy $\varepsilon = \hbar v$, where v is a central frequency of the radiation spectrum. The "corpuscular" properties of *the C-photon* reveal themselves only at the moment of detection. But there are quantum optical effects: the essential quantum effects that have no classical analogues. Such effects cannot be described in the framework of the semi classical model based on Maxwell's equations.

2) ***The M-photon*** is a hypothetical elementary particle of the light field generating an impulse at the output of the photodetector. Although there is no more rigorous definition of *the M-photon* in the framework of any consistent theory, the photon as a particle (with the wave properties characteristic of the elementary particles) is used in various optical studies where an attempt is made to go beyond the framework of the Copenhagen interpretation. Here, it is assumed that any radiation field consists of a set of almost independent *M-photons* with definite *a priori* features to be revealed after a time.

It is interesting that the first corpuscular models of the light field consisting of the elementary particles, each with energy $\hbar v$ where v is the radiation frequency, were developed after A. Compton's experiments on X-ray scattering (1922). The observed change in the frequency of the scattered radiation was explained by the elastic collision of an electron and a particle possessing energy $\hbar v$ and momentum $p = \varepsilon / c$ In 1926, G.H. Lewis called this particle a photon.

Note, the model of the C-photon, as a classical wave packet, does not contradict the model of the M-photon as a hypothetical elementary particle if propagation of light is a nonlinear process in a medium like a soliton (the density inside a soliton can be different than that of the surrounding medium).

3) *The Q-photon* is an objective entity corresponding to the Fock state of the light field with $n = 1$ or a superposition of such states with nearly equal energies. This definition can be made in terms of the standard quantum theory of light. However, the statement that "light consists of photons" suggesting the definite number n of such constituent elements of light does not make any sense in the standard quantum theory because the field has no definite n before measurement. Of course, a problem of interpreting the quantum formalism still remains. The Copenhagen interpretation forbids asking nature "idle" questions, that is, it has a pragmatic tint. In the framework of this interpretation "A photon can be called a photon if only it is a detected photon". Only investigating the characteristics of the pure or combined state of the field is permitted.

There is a case where all the above mentioned types of photon appear consistent: when the light field is in the one-photon state (photon in the pure state). In this case a priori properties of the photon can be discussed.

2. Alignment of the light's velocity to the known value C near the Earth's surface

In 1908 Walter von Ritz suggested that $\bar{c}$ was the velocity of light with respect to the source and the classical law of composition of velocities was valid for the case of the moving source (the so-called Ritz ballistic hypothesis) [2]. Under this assumption the aberration of starlight, the results of the famous Michelson-Morley experiment, and those of most other experiments aimed at detecting the aether wind come into agreement with each other.

However, the experiment performed at CERN, Geneva, in 1964 was considered to be the most convincing evidence against the Ritz theory.

In this experiment the speed of 6 GeV photons produced in the decay of very energetic neutral pions was measured by time-of-flight over paths up to 80 meters in length. The pions were produced by the bombardment of a beryllium target with 19.2 GeV protons having speeds (inferred from the measured speeds of charged pions produced in the same bombardment) of $0.99975\,c\ldots$ [3]. Within experimental error it was found that the speed of the photons emitted by the extremely rapidly moving source was equal to c. If the observed speed is written as $c' = c + ku$, where u is the speed of the source, the experiment showed

$$k = (0 \pm 1.3) \cdot 10^{-4} \qquad (1)$$

The following three different hypothesis agrees with the CERN experiment.

The first hypothesis: **the speed of the photon equals c, when measured with respect to the Earth, and is independent of the velocity of the source.** Indeed, the experiment performed at CERN could be explained by this hypothesis in the frame of the model of 3-dimensional Euclidean space, but it brings other problems that take us back to the "crisis in physics".

The second hypothesis – this **hypothesis** is the basis of the special relativity: **the speed of light has the same value c with respect to any inertial frame of reference.** That is from the standpoint of SR the speed of an emitted photon measured with respect to an inertial frame of reference associated with a moving source is equal to c, (which also **agrees with the Ritz hypothesis.**) On the other hand, from SR it also follows that the speed of a photon measured with respect to Earth is also equal to c, which agrees with the experiment performed at CERN. SR provides an explanation of the above two statements by discarding the classical law of composition of velocities and the hypothesis of aether as a preferred reference system, and introducing a model of four-dimensional pseudo-Euclidian space.

The third hypothesis: **Ritz ballistic hypothesis plus the modified extinction theorem.** It can be concluded from the fact that relativistic kinematics correctly describes the results of certain optical experiments that in the four-dimensional kinematic formalism of special relativity there are dynamics 'hidden' in the geometry of space. This idea was first put forward by E.L. Fainberg in 1997 [4].

In other words, it is possible to explain optical experiments remaining in the framework of the model of the three-dimensional Euclidean space and the classical law of composition of velocities. But in this case we have to assume that **the speed of light (photon) can change within the same real physical frames of reference** (for example, leaving a source with a velocity $\vec{c} + \vec{u}$ where $\vec{u}$ is a velocity of the source with respect to the Earth, the speed of photon acquires the value c near the Earth's surface as is observed in experiments.)

Note that there were earlier attempts to prove the consistency of Ritz's theory using the extinction effect {Ewald (1912), Oseen (1915), Fox (1962)}. According to this effect when an electromagnetic wave is incident on a homogeneous medium it is extinguished inside the medium in the process of interaction and is replaced by a wave propagated in the medium with a velocity different from that of the incident.

The experiment performed at CERN, however, demonstrates that the change of the photon's speed to the value c would have to occur even in vacuum. Because of that the extinction theorem was proven to be wrong and was not accepted anymore by conventional physics.

From our point of view, the idea that the velocity of a photon changes its value to c near the Earth's surface has not exhausted itself. However the nature of that process is, probably, closely connected with the interaction of the photon with the physical fields associated with the Earth. We will come back to this discussion in chapter 5. Here we would like to emphasize that the third hypothesis is in agreement with the majority of optical experiments. We will show below that the transverse Doppler Effect can also be explained on the basis of the third hypothesis. Note that the observation of the transverse Doppler Effect is still one of the main arguments in support of relativity kinematics.

Using CERN let us estimate the length of the path l on which the speed of the photon emitted by the moving source remains $\vec{c} + \vec{u}$ in compliance with the Ritz theory, where $\vec{u}$ is the speed of the source with respect to the Earth. If we take the speed of the source $\vec{u}$ to be approximately equal to $\vec{c}$ and set, according to the formula (1), $k = 10^{-4}$ for the experimental error then, within the accuracy of the experiment, the average velocity of the photons emitted by the moving source is $c' = c(1 + 10^{-4})$ with respect to the Earth. Then the length of the path l on which the velocity of the photon remains $2c$ (that is remains equal to c with respect to the source) would be $l = 1.6 \times 10^{-2}$ (assuming that all the photons traveled 80 meters). This is a large distance even for daylight photons (a photon with energy 0.25 **eV** has wavelength 0.5×10^{-6} m, that is on the path of length l we have~3×10^4 wavelengths).

3. The derivation of the formula for the transverse and longitudinal Doppler Effect using the classical mechanics law of composition of velocities

Below, the equation for the transverse and longitudinal Doppler Effect is derived for the case of a photon in the pure state. In this case the properties of the photon can be discussed: its energy, momentum, mass, polarization. We assume that the classical law of composition of velocities and the law of conservation of energy and momentum are valid.

Case 1: Suppose that a source of light is at rest with respect to the Earth, and an observer is moving with a constant speed $-\overline{u}$ relative to the Earth. In the frame of reference of the Earth, the speed of the photon emitted by the source is equal to c and there is no reason why it should change in the observer's frame of reference prior to interaction of the photon and the detector.

We will work in the frame of reference of the observer. In the observer's frame of reference a source of light with mass M is moving with velocity $\overline{u}$ (Fig.1). The energy of the source is composed of kinetic energy $Mu^2 / 2$ and internal energy E_0 of the excited atoms. Denote by E' the internal energy of the source after the photon is emitted. In addition the source undergoes recoil due to emission: its speed gains an increment of $\overline{u}' - \overline{u}$ (where $\overline{u}'$ is the speed of the source after emission of the photon). From the laws of conservation of energy and momentum for the photon and the source respectively, it follows that

$$\frac{M\,u^2}{2} + E_0 = \frac{(M - m_0)\,(u')^2}{2} + E' + E_{ph} \qquad (2)$$

$$M\,\overline{u} = (M - m_0)\,\overline{u}\,' + m_0\,\overline{w} \qquad (3)$$

where m_0 is the mass carried away by the photon emitted with speed c with respect to the source, E_{ph} is the photon energy in the observer's frame of reference, and $\overline{w} = \overline{c} + \overline{u}$ is the photon velocity in the same frame.

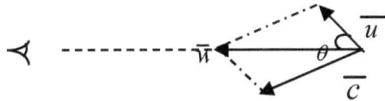

Fig. 1

Note that the vector $\overline{w}$ is directed towards the observer From Eq. (3) we obtain for $\overline{u}\,'$:

$$\overline{u}\,' = \frac{M\,\overline{u} - m_0(\overline{c} + \overline{u})}{M - m_0} \qquad (4)$$

After emission of the photon, the internal energy of the atom is decreased by the amount $h\nu_0$, where ν_0 is the natural frequency of the atom, that is $E_0 - E' = h\nu_0$. Taking this along with Eq. (4) into account, Eq. (2) can be expressed as follows:

$$\begin{aligned} E_{ph} - h\nu_0 &= \frac{M u^2}{2} - \frac{(Mu - m_0(\overline{c} + \overline{u}))^2}{2(M - m_0)} = \\ &= \frac{m_0 u^2 + 2 m_0 (\overline{u} \cdot \overline{c}) - m_0^2 (\overline{c} + \overline{u})^2 / M}{2(1 - m_0 / M)} \end{aligned} \qquad (5)$$

If the mass M of the source is much greater than that of a photon, the terms containing m_0 / M may be ignored. In this approximation, Eq. (5) takes the form:

$$E_{ph} = h\nu_0 + m_0(\overline{u} \cdot \overline{c}) + m_0 u^2 / 2 \qquad (6)$$

Using the relation $m_0 c^2 = h\nu_0$ (note that this is not a consequence of special relativity), Eq. (6) can be represented in two equivalent forms:

$$E_{ph} = h\nu_0 \left(1 + \frac{(\overline{u} \cdot \overline{w})}{c^2} - \frac{u^2}{2c^2} \right) = \frac{m_0 w^2}{2} + \frac{h\nu_0}{2} \qquad (7)$$

where

$$w^2 = c^2 - u^2 + 2uw\cos\theta \qquad (8)$$

Here θ is the angle between the velocity of the source and the direction from the source to the observer, i.e. the angle between vectors $\overline{u}$ and $\overline{w}$.

Consider the special case $u = 0$. In this case Eq. (7) implies:

$$h\nu_0 = \frac{m_0 c^2}{2} + \frac{h\nu_0}{2} \qquad (9)$$

A very important result follows from Eq. (7) and Eq. (9): **the energy of a photon, as an entity with mass m_0, can be represented as a sum of two terms, the first being the kinetic energy of the center of mass, in which we assume all of the photon's mass is concentrated; the second being the energy associated with the motion about the center of mass, which is characteristic of the photon's intrinsic degrees of freedom.** Formula (9) was obtained by L.Boldyreva and N. Sotina in 1999. [5].

It is experimentally established that the absorption of light occurs in a quanta of energy $h\nu$, where ν is the detected frequency. Assume that all the energy of the photon E_{ph} is equal to the energy detected by a measuring device, that is $h\nu$ (this assumption is no different than that of conventional physics). Under this assumption, from equation (6) we obtain

$$\nu = \nu_0 \left(1 + \frac{(\overline{u} \cdot \overline{w})}{c^2} - \frac{u^2}{2c^2} \right) \qquad (10)$$

If $\overline{u} \perp \overline{w}$, that is, $(\overline{u} \cdot \overline{w}) = 0$, then the expression for the transverse Doppler effect follows from Eq. (11):

$$\nu = \nu_0 \left(1 - \frac{\beta^2}{2} \right) \qquad (11)$$

Using Eq. (10) and Eq. (8) we obtain the detected frequency of the photon for any value of θ :

$$\begin{aligned} \nu &= \nu_0 \left(1 + \beta \frac{w}{c} \cos\theta - \frac{\beta^2}{2} \right) = \\ &= \nu_0 \left(1 + \beta \cos\theta - \frac{\beta^2}{2} + \beta^2 \cos^2\theta + O(\beta^3) \right) \end{aligned} \qquad (12)$$

Eq. (12) agrees, to within an accuracy of $\beta^2 = (u/c)^2$ inclusively, with that of describing the Doppler Effect in special relativity.

Note also that as follows from Eq. (12), the frequency remains the same ($v = v_0$) in the following two cases: 1) when the relative speed of the source is zero ($u = 0$), and 2) when $u = 2c\cos\theta$. In these cases $w = c$ and, consequently, the total energy of the photon is the same in both frames of reference. The relativistic equation for the Doppler Effect also has two solutions when the frequency of light remains unchanging, however, in SR the second solution agrees with our solution only approximately (with an accuracy of β^2 inclusively) and does not have an obvious physical interpretation. The fact that in our consideration the second solution is the **exact** solution of Eq. (12), and has a simple physical interpretation is an additional argument in favor of the theory developed in this work.

Where is "hidden dynamics" here? In our derivation we take the energy of the absorbed photon to be hv. In agreement with conventional physics let us use the expression

$$hv = mc^2 \qquad (13)$$

for the energy of the absorbed photon. It follows from this formula that the mass of the photon changes as the value of the velocity changes to the value c in the vicinity of the detector. Then the change of the momentum near the detector is

$$\Delta \vec{k} = m\vec{c} - m_0(\vec{c} + \vec{u}) \qquad (14)$$

Here $\Delta \vec{k}$ equals to the impulse of external forces. In cases when the angle θ between the velocity of the source and the direction from the source to the observer is 0 or π formula (14) gives

$$\Delta k = mc - m_0(c \pm u) = \frac{\hbar}{c}[v - v_0(1 \pm \beta)] = m_0 c \frac{\beta^2}{2} \qquad (15)$$

It can be seen from Eq. (15) **in the first approximation by β** that $\Delta k = 0$. Therefore, in the first approximation by β the **hidden dynamics is in the change of the photon's speed which occurs at the expenses of the change of its mass. That is, as the speed of the photon increases to the value c its mass decreases and vise versa, as the speed of the photon decreases to the value c its mass increase in the vicinity of the detector.** (Note, that here we expand Ritz hypothesis: the speed $\vec{c}$ is the speed of light with respect to both the source, and the detector).

Case 2: Now suppose that an observer is at rest with respect to the Earth, and a source is moving with constant speed $\bar{u}$ relative to the observer. In this case the emitted photon has the speed c and energy hv_0 in the frame of reference of the source. The speed of the photon with respect to the Earth is different from c when the photon is emitted but changes to the value c at the vicinity of the source. In this case Eq. (13) for the Doppler Effect remains valid, however, v in this equation is the photon's frequency with respect to the Earth.

Conclusion. In the above it was proven that the relativistic equation for the Doppler Effect can be obtained in the framework of the model of the three-dimensional Euclidean space using the classical laws of conservation of energy.

From the law of conservation of energy it follows that **1)** the energy of a photon, as an entity with mass m_0 can be represented with two terms: the first is the kinetic energy of the center of mass; the second is the energy associated with the motion about the centre of mass; **2)** In the process of absorption of a photon by a moving detector all energy of the arriving photon is absorbed by an atom.

4. Light Curve for Eclipsing Binary Stars

We have spoken so far about the change of the photon's velocity near the Earth. A question arises: are there observations in outer space which can be explained by "hidden dynamics "? That is, observations that the photon's speed changes in vacuum (without loss of energy) within the same frame of reference? The answer is yes: the astronomical observations of the motion of binary stars.

At one time (1913) astronomical observations of binary stars was the single objection to the ballistic hypothesis of Ritz. It is generally accepted that the paper of W. de Sitter [6] put an end to the Ritz idea. In his work W. de Sitter pointed out, that if one follows the hypothesis that the speed $\bar{c}$ is the speed of light with respect to each of the stars and the classical law of composition of velocities is valid, then light, emitted simultaneously from each star reaches the Earth at different moments. As a result an observer on Earth can observe the discrepancies with Kepler's laws.

De Sitter based his reasoning, however, on a hypothesis that the speed of light is unaffected during its journey to Earth. Our assumption that « hidden dynamics» exists allows bringing the hypothesis of Ritz into agreement with the observations of the motion of binary stars. Moveover, the observations of binary stars can help to estimate at what distances from the stars the speeds of the photons emitted from each of the stars in binary system become equal.

In our analysis we use the same assumptions as SR: speed of light equals c with respect to each star and it also equals c with respect to an observer on Earth. However, from our point of view the key is not in the relativistic law for velocity addition but in a a real change of the speed of light that take place as it propagates in space, that is in existence of 'hidden'dynamics that manifests itself in the change of the light speed (in vacuum) without energy loss.

Consider the case of eclipsing binary stars, a system of two stars A and B, whose plane of orbit lies in the line of sight of the observer. According to our hypothesis the speeds of photons emitted by star A are equal to c with respect to that star, and similarly the speeds of photons emitted by star B are equal to c with respect to star B (Fig.2). Denote as $\overline{u}$ the velocity of a star in a binary system about the common center of mass (for simplicity considers $\overline{u}$ being the same for both stars). Due to the motion of the stars the speeds of photons moving in the direction of the line-of-sight of the observer should be different. After some time, however, the speeds of the two sets of photons can 'equalize' and acquire the same value c, for example, due to their passing near another celestial body.

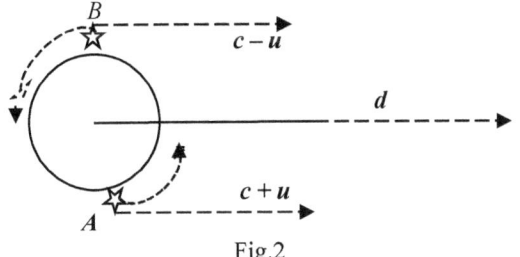

Fig.2

Let d be the distance at which speeds of photons equalize. This way, the light curve plotted by the observer located at a distance d from the binary system (call this point M) is the same as the light curve plotted by the observer on Earth (because the photons travel further with the same speed).

The relationship between the current time t and the time of the photon's arrival at point M (for both stars A and B) is given by the following equation:

$$\frac{d}{c(1+\beta\cos\omega t)}+t=\tau_1 \qquad (16)$$

for the photon emitted by star A, and

$$\frac{d}{c(1-\beta\cos\omega t)}+t=\tau_2 \qquad (17)$$

for the photon emitted by star B.

Let ω indicate the angular speed of the stars' orbital motion about the common center of mass, $\omega = 2\pi/T$ where T is the orbital period, and $\beta=u/c$. The position of the stars at the initial moment of time ($t = 0$) is shown in Fig.2.

Assume that the number of photons emitted per unit time n is the same for both stars. Let μ_1 be the number of photons per unit time arriving at the point M from the star A, and μ_2 be the number of photons per unit time arriving at the point M from the star B. In the time interval Δt each star emits $n\Delta t$ photons. The number of photons arriving at point M is therefore $\mu_1(\tau_1)\Delta\tau_1$ and $\mu_2(\tau_2)\Delta\tau_2$ respectively. Then for star A we have

$$n\Delta t = \mu_1(\tau_1)\Delta\tau_1 \approx \mu_1(\tau_1)\frac{d\tau_1}{dt}\Delta t \qquad (18)$$

and for star B:

$$n\Delta t = \mu_2(\tau_2)\Delta\tau_2 \approx \mu_2(\tau_2)\frac{d\tau_2}{dt}\Delta t \qquad (19)$$

where $d\tau_1/dt$ can be found from Eq. (16) as

$$\frac{d\tau_1}{dt}=1+\frac{d\beta\omega\sin\omega t}{c(1+\beta\cos\omega t)^2}, \qquad (20)$$

and $d\tau_2/dt$ can be found from Eq. (18) as

$$\frac{d\tau_2}{dt}=1-\frac{d\beta\omega\sin\omega t}{c(1-\beta\cos\omega t)^2} \qquad (21)$$

We are studying the change in light intensity in the frame of point M. Thus, we have to substitute t with τ_1 and τ_2 in Eq. (20) and Eq. (21) respectively. According to Eq. (18) the relative density of photons arriving at point M from star A is

$$\frac{\mu_1(\tau_1)}{n}=1/\frac{d\tau_1}{dt}(t\to\tau_1) \qquad (22)$$

According to Eq. (21) the relative density of photons arriving at point M from star B is:

$$\frac{\mu_2(\tau_2)}{n} = 1 / \frac{d\tau_2}{dt} (t \to \tau_2) \qquad (23)$$

So, the total relative density s of photons arriving at point M is as follows:

$$s = \frac{\mu_1(\tau_1) + \mu_2(\tau_2)}{n} = \frac{1}{d\tau_1 / dt} + \frac{1}{d\tau_2 / dt} \qquad (24)$$

From the viewpoint of SR, $s = 2$, and the graph of s versus time (Fig.3) should be constant between eclipses. In our case the graph of the function s given by Eq.(25) shows that the curve, which represents the relative photon density $s=s\,(t/T)$ as measured at the point M, is a periodic function with the period $T/2$ (where T is the orbital period of the star system) (Fig.3).

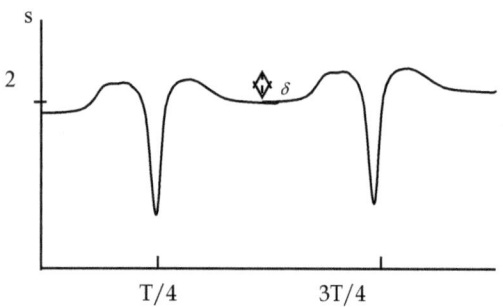

Fig. 3. *Light curve of an eclipsing binary system*

The variations δ from $s = 2$ depends on the distance d from the star to the point M (the point where the photons' speeds equalize). Using data for the binary system WW Aurigae, we estimat that at a distance $d = 10$ AU, $\delta = 8.463 \times 10^{-8}$, and for $d = 1000$ AU, $\delta = 6.113 \times 10^{-5}$. In the case of WW Aurigae δ is small, and is probably not detectable in observations.

Thus a light curve plotted on the basis of SR is different than the curve plotted on the basis of our theory. Light curves showing uneven brightness, however, are often observed. Besides the drops in intensity due to eclipses, there are observed deviations from constant values in the regions of light curves between eclipses. Astronomers have different explanations for these variations, some of which are quite obviously contrived. This topic clearly requires further study to arrive at a credible resolution. And yet the new results of the observation of binary stars might provide new arguments in favor of the existence of "hidden dynamics".

5. *Some of the hypotheses*

It can be assumed, that propagation of light is a process in the aether and the speed of light is equal to $\overline{c}$ with respect to the aether. If the aether has mass, that is, it has gravitational properties it should be attracted to any other mass (to the Earth, to the Sun, to a star, etc.), thereby covering it. In view of the above a natural "candidate" for the role of the aether is dark matter. Dark matter is known to constitute about 85 % of the total matter in the universe, while ordinary matter makes only about 15 % (in percentage of the total mass-energy of the universe: ordinary matter is about 5%, dark matter is 27%, and dark energy is about 68%). At the present time the nature of the dark matter is unknown. Large astronomical searches for gravitational microlensing and detailed analysis of the small irregularities in the cosmic microwave background show that most of dark matter has a non-baryonic nature (meaning it does not consist of electrons, protons and neutrons like ordinary matter).

If dark matter is the aether, then the fact that dark matter does not interact with light or other electromagnetic radiation has a very simple explanation. Propagation of light is a process in the aether, and, therefore, light does not interact with it as it does with external matter.

As is known, the classical argument against the model of the aether that is carried with the Earth is the observation of aberration of light. Let us recall that the aberration of starlight is the apparent displacement in the positions of stars caused by the yearly motion of the Earth in its orbit (the yearly aberration of light). The phenomenon was discovered in 1727 by J. Bradley observing stars through a telescope. Bradley interpreted the discovery from the point of view of Newton's corpuscular theory of light. From the wave point of view it followed that the aether surrounding the Earth does not share the Earth's motion.

These problems was the one that brought the "crisis in physics" that was settled with the development of SR at the end of the 19th, beginning of 20th century. However, it would be sufficient to assume that the light wave is not a regular wave, like sound, but rather is like a soliton, and the problem of the aberration of starlight would be solved. As is known solitons are localized within a region of a medium where they propagate and can possess an inertia properties. The hypothesis, that light propagates like a soliton in the aether explains many optical experiments, however, some of the phenomena cannot be explained on the basis of this hypothesis only. (for example, the transverse Doppler Effect). From our point of view, the search for new approaches to the explanation of these phenomena should be not in direction of the construction of the space-time geometrical models but rather in the study of the properties of physical vacuum.

The aether in our opinion is in many ways similar to superfluid ^{3}He [7,8]. The superfluid properties of the aether (zero viscosity while in motion) could explain the observed nondissipative motion of celestial bodies in space. A vortex-wave process in the medium with the properties of the superfluid of ^{3}He has unique properties that could explain the rectilinear propagation of light, the existence of the photon's mass, as well as the electrical and quantum properties of a photon. Indeed the following phenomena are observed in the superfluid ^{3}He:

1) quantization of the angular momentum in vortices;

2) inertial properties of vortices (in the cores of vortices of superfluid ^{3}He-B phase transitions may take place which may lead to a change in the inertial properties of the medium in the vortex compared to the inertial properties of the superfluid in the rest of the volume);

3) solitons, named "hedgehogs" (the Barnett effect in superfluid ^{3}He-B leads to a possibility of generating vortices which terminate in the superfluid due to the complete transfer of the vortex angular momentum to the orbital angular momenta and spins of the particles constituting the vortex);

4) magnetization of the cores of vortices along the vortex axis (this means that the spins of ^{3}He atom,ms are oriented along the vortex axis);

5) electric polarization of vortices (the vortices in superfluid ^{3}He-B are electric dipoles).

6) in superfluid ^{3}He-B structures like the homogeneous precessing domain are observed, where all spins of the fluid particles precess with the same frequency and phase.

In the article "The 'hidden variables' in quantum mechanics. The precession of the electron's spin in an atom", published in this edition of the NPA proceedings and also in work [9], it is proven from the Schrödinger equation on the basis of the causal interpretation of quantum mechanics that an electron's spin in an atom precesses. In this case from the standpoint of the model of the superfluid aether the motion of an electron in an atom creates structures in the aether. If we consider these structures as quasi-particles of the superfluid physical vacuum that have spin it can be shown that the natural frequencies of the atom are the frequencies of the precession of the quasi-particle's spin. It follows from this hypothesis that the spin polarization of the superfluid aether must have a significant effect on the formation of the photon and on its velocity..

Acknowledgment. We would like to thank Anatoly Sukhorukov for his contribution to calculation of the light curve for binary stars.

References

1. D.N. Klyshko, *Physics Uspekhi*, **164 (11)**, 1187 (1994) (in Russian).

2. W. Ritz, Annales de Chimie et de physique, **8**, 145 (1908).

3. T.Alvager & J.M. Bailey *Phys. Letters*, **12**, 260 (1964).

4. Fainberg E.L., *Physics Uspekhi*, **167 (11)**, 455, (1997) (in Russian).

5. L. B. Boldyreva & N. B. Sotina, A Theory of Light without Special Relativity? (Moscow: Logos, 1999).

6. W. Sitter, *Physikalische Zeitschrift*, **XIV**, 429 (1908).

7. N. Sotina, Reports of the 16th RCCNT&BL Dagomys, city of Sochi, June, 2009, Moscow 179 (2010) (in Russian).

8. L. Boldyreva & N. Sotina, *Physics Essays*, **5**, 510 (1992)

9. N. Sotina, *Physics Essays*, **27**, 321 (2014).

THE LANGUAGE OF POWER
PETER MARQUARDT

Stolberger Straße 111, D-50933 Köln GERMANY
e-mail: marquardtp@gmail.com

Covering a huge range of topics, physics is influenced in various ways not always helpful to its role as a natural science in its own right. Particularly the misuse of mathematics facilitated to establish dogmas, making physics the battlefield of what may be called the *language of power* (LOP). It is this manifold voice of mainstream control over physics against justified criticism that we have to be aware of. The present essay exemplifies some ways of manipulating physics: Formulas, fancy names, ill-defined concepts, authority, fame, propaganda, *gedanken* experiments, "correct" results, the "*heureka effect*", or the hasty acceptance of "established knowledge". Investigating the traces of LOP is good for some surprises and gives us a valuable lesson for our everyday lives beyond physics.

Dedicated to all members of our "dissidents' family" who have passed away.

"Language of Power"?

This strange language is more familiar to us than meets the eye.

Lawyers earn their money with the language of power (LOP). It is their way of lending their voice to their clients who would be defenseless facing the jungle of juridical laws. Here, the aim of LOP obviously is *manipulation*. The *rules* set up by the judiciary leave enough room for interpretation and alterations. Giving these rules the status of a "law" already constitutes an example of a LOP officially used in our everyday lives. A natural scientist has (or should have) a distinctly different view of what constitutes a "law". In its strict sense, a law is a principle that allows no exception and cannot be violated. How many "laws" then do we have in physics? Most of the "laws" in physics, usually expressed in terms of mathematics, suffer from lots of exceptions. Yet they are treated like untouchable verities. This brings us to the problem of LOP in natural science, physics in particular. Usually, physics is considered an exact science founded on its own strict logic that would not yield to interpretation or amendments or adjustments. Math, above all, is the art of abstractions. Math knows no other authority besides its own logic; but it seems physics doesn't. Math and physics are not as close together as is generally believed. The scope of physics, unlike that of math, reaches far beyond abstractions. Yet physics cannot do without abstractions, either, which makes it vulnerable to math. The vulnerability to math already raises some suspicion that some kind of LOP may be active in physics. Upon closer inspection of many problems, we realize that a LOP invades the official treatment of physics much deeper than we would like to admit. In fact, the effect of LOP comes in many disguises. Fortunately, we can stay very simple in visualizing these disguises. The present essay exemplifies some aspects of LOP active in physics and thus intends to be an appetizer for the search of more. Any distinction of the various activities of the LOP in physics is somewhat artificial as they are interwoven in a tricky way. Math is a good start to enter the subject.

Mighty Math: Formulas, Formulas...

Math and physics are two distinctly different worlds, each a science in its own right. Yet the mathematical LOP rules in physics, surprisingly often to the disadvantage of physics. If it is not the faithful servant, math can be disappointingly destructive, leaving physics rather helpless. A "correct" result often leads astray. A formula has to be backed up by an analysis in terms of solid physical principles. Unfortunately, some prominent theories owe their fame to a purely mathematical procedure.

Note: *A correct result is not sufficient to prove a physical theory right.*

Some formulas became so famous they don't require any explanation as to the meaning of their symbols. The associated names are sufficient: *Planck* Blackbody Radiation, *Newton* Gravitation, *Lorentz* Transformation, *Boltzmann* Entropy, *Maxwell* equations - you name them.

One of them, *Mass-Energy Equivalence*, deserves special attention. It, in particular, owns its fame to propaganda, although (or because?) it is an as simple as can be formula. Its simplicity contrasts the physics it claims to represent. Books have been written about these very few symbols, $E = mc^2$. They became the trademark of modern physics, a graffiti for genius, and, erroneously, the flag for relativity. Excited by *Einstein*'s fame, the public jumped to a status of "experts" on energy and the free trade between matter and energy became the big hope to solve all energy problems. The conversion of mass to energy is *not* what the formula says. Dimensions tell us that mass and energy are different. Energy is *context dependent* and can only be defined *in connection with a complex interacting system*. That also puts c^2 in the category of context parameters. Classical mechanics teaches us that (velocity)^2 terms are dynamic potentials as is known from $v^2/2$. Care should be taken not to confuse dynamics with kinematics and thus carelessly import energy into the ambiguous world of relative velocities where c^2 has no place. A lonesome non-interacting ("free") particle just does not have a velocity all by itself and it would be foolish to speak of "self-energy", just by attaching the c^2 factor to its mass. The potentials to be associated with *bound* particles like those in an atomic nucleus do not show up in mc^2. The energy liberated by fusion or fission was stored in the original configuration of the system. If this is not explicitly respected, how can we be sure that the "mass defect" is not calculated from E/c^2 in a vicious cycle to "prove" $E = mc^2$? If left to math, that notorious c^2 remains a stranger unless backed up by at least an attempt to give it some physical significance. One promising aspect might be to view the mysterious c^2 as a ubiquitous background potential which makes $E = mc^2$ look more innocent than sensational. Potentials are a successful and consistent way to take interactions into account. It is a bad habit to neglect constant potentials and make them an arbitrary dimensionless zero. And the paradigm "only conservative forces have a potential" is a premature mathematical statement and should be formulated more precisely: "Conservative forces have an *analytical* potential". Friction certainly works on the same principle as all forces do, whether we can formulate the potentials or not. Forces are more obvious than energy, a late comer in physics. Forces receive more attention and the gradients of static and dynamic potentials are erroneously regarded as more fundamental than the potentials themselves. Potentials play an important role in dynamics. They establish the context in an interacting system.

The c^2 potential is constant. For dynamics, we may introduce a velocity term ("gamma factor") $\gamma(v)$. Math does not care whether we arrive at $\gamma = (1 - v^2/c^2)^{-1/2}$ by means of kinematics (*special relativity* and transformation) or dynamics (*neomechanics* and

energy). In both cases, the formula looks identical, the meanings are fundamentally different, v being the unique absolute velocity in neomechanics, the proper stage where c^2 belongs. It is benevolent (or "blind") math like this that helped questionable theories survive in spite of their fundamental flaws.

Planck arrived at his formula in search for the best mathematical fit between two limiting cases (see below). The physical analysis of thermal radiation started with *entropy*, a true thermodynamic concept that certainly doesn't apply to a single particle. Nothing compels us to interpret radiation in terms of single photons. *Planck* scaled down some constants. Dividing the gas constant (used by *Boltzmann*) by *Avogadro*'s number he arrived at what became known as "*Boltzmann*'s constant" on the molecular level. Does this make it a "quantum of entropy"?

Theoreticians tend to overdo their mathematical treatment. A bad case is putting elementary constants $c=e=h = 1$, wiping out all traces of the all-important physical dimensions and emphasizing a probability factor instead. Theoreticians also like to introduce the imaginary unit. This is sometimes quite helpful, e.g. in the representation of alternating current circuits, but in quantum theory it leads away from physics. Of course, math can come up with astonishing precision so as to confirm a result to the nth decimal in praise of a successful collaboration between precise theory and experiment. Take the anomalous magnetic moment of an electron. Great achievement or propaganda? Hard to decide for an outsider.

Propaganda, Propaganda

Propaganda, especially the one-sided or biased kind, is a first step in the creation of a dogma. Permanent propaganda in praise of a theory, resulting from "heureka" and success, is always suspicious. It intends to put critical minds to sleep and puts a "forever right" label on the theory.. The celebrated cult theories in physics benefit from propaganda, both relativities, Copenhagen quantum theory, Big Bang, quantum electrodynamics, and, sadly, also *Maxwell*'s theory.

Propaganda is responsible for the selective choice of experimental results. Take the most famous "experiment that failed": *Michelson-Morley* (MM)! It was a brilliant experiment that did *not* fail at all; the by hook or by crook interpretation of its officially accepted null result did. Non-zero fringe shifts (see e.g. the work by *Héctor Múnera*) are annoying to a certain cult theory. Instead of searching for an analysis combining the different conditions that lead to *all* kinds of fringe shifts (including *Sagnac*'s results with his rotating interferometer!), propaganda prefers to stay biased.

Propaganda is responsible for the fuss made about the "unexplained rest" of 42 arc-seconds/century of Mercury's perihelion precession. This deliberately belittles the classical lion's share of almost all of the total 5600 arc-seconds/century which is actually observed.

Propaganda is responsible for the triumphant progress of the single photon philosophy. It started with *Einstein*'s naïve interpretation of the photo effect (the "first application of *h*" which made *Planck* support young *Einstein* in gratitude*)* that is hoorayed to this day. The biased interpretation of *Planck*'s formula in terms of statistically independent single photons has become a dogma. But even thermal sources emit coherent radiation. Coherence is not the exclusive trademark of a laser. A laser has an extremely high degree of coherence. Important evidence by low intensity interference that breaks down below a threshold demonstrating the emission of *coherent photon bunches* (check the experiments by *Yu. P. Dontsov* and *A. I. Baz*, and by *Emilio Panarella* on internet) is widely disregarded.

Propaganda is responsible for the exclusive use of the *Lorentz* force $e[\mathbf{v}\times\mathbf{B}]$ on a charge e with velocity $\mathbf{v}$ in a magnetic field $\mathbf{B}$. It is advisable to use *potentials* and not their gradients in dynamics. If the vector potential, $\mathbf{A}$, is multiplied by $\mathbf{v}$ we get the scalar potential $(\mathbf{A}\mathbf{v})$. *Karl Schwarzschild* termed it "electrokinetic potential". The gradient of $(\mathbf{A}\mathbf{v})$ delivers four force terms. *Lorentz*'s is only one of them.

Propaganda is responsible for the official view of light bending in a gravity gradient or of the redshifted radiation from deep space. Mainstreamers do not discuss alternatives. You will easily identify more one-sided propaganda easily recognized by its notorious repetitions.

Linguistic Aspects –Laws or Rules or Just Formulas?

Languages are sometimes hard to use in a strictly logical way, because they lack logic to some degree. In natural science, this may damage its claims to be an exact science.

Names, especially those we are used to, can put a wrong label on what we want to say. We speak of "atoms" which since the rise of nuclear physics isn't true anymore in its original meaning. We are using thermometers (θερμοτης = "heat") to measure temperature, not heat. Yet we don't call them temperameters. The usage of language often is ambiguous and may facilitate to establish the LOP as the only voice of science.

We know why calling a rule a "law" (that sounds more impressing) stabilizes the hierarchy of a system in power. This kind of power, however, is not helpful to a science. With "laws" and "principles", physicists have to be cautious, especially when a formula does not specify whether we have to do with a law, a rule or just a formula. Any two-body interaction becomes an approximation, requiring sophisticated math in more complex systems, starting with as few as 3 objects. This makes *Newton*'s gravitation and the *Cavendish - Coulomb* force (see below) between two charges a "law" strictly applicable to just 2 interacting objects. The tacit extrapolation of $1/r^2$ laws to more complex systems leads to well-known paradoxes like those named after *Olbers* ("why isn't the Universe bright like daylight?") and *von Seeliger* ("why isn't Earth been torn apart due to the attraction by all other masses?"). These paradoxes can be resolved by a slight attenuation of the Newton $1/r^2$ law.

Energy conservation and its early predecessors (*Newton* formulated his three Principles before "energy" became recognized) are *laws* in their own right. Energy conservation is a principle too general to be squeezed into a single formula. Energy conservation is a mighty help when it comes to judge a formula. A violation of *Newton*'s Third Law (III) is a serious matter. That should be the number one criterion to accept or reject a formula. The *Grassmann-Biot-Savart* force violates *Newton* III except in a rigid closed current loop only and it admits transverse forces only. This suffices to make it practically useless. Notwithstanding, mainstreamers consider it the one and only force between currents, ignoring important experimental evidence. Longitudinal forces are excluded from mainstream electrodynamics. *Ampère*'s force law comprises all forces and is in accordance with *Newton* III. *Ampère*'s repulsive longitudinal forces between collinear currents have been verified with the *Ampère* bridge, demonstrating the failure of the officially recognized formula by *Grassmann-Biot-Savart,* thus disqualifying it. On closer inspection, *Faraday*'s induction is not a law, it is a relationship between two concomitant effects, both produced by a vector potential *A* changing with time. We note that *Faraday* and *Grassmann-Biot-Savart* are the basis of two of *Maxwell*'s equations...

Language can be a mighty illusionist. When relativists face severe inconsistencies in their line of argumentation (see the infamous "twins"), they call it a "paradox" where they should clearly put "absurdity". A paradox in its proper meaning is an *apparent* contradiction that *can* be resolved. The absurdity of relativity cannot be removed by inventing more reference frames.

Language sneaks concepts into theories that are strangers in physics ("no *information* can travel faster than c" or "*probability* waves"). The fantasy of some theorists is remarkable. Some of them invented the "watchdog effect", a mechanism intended to keep an otherwise spreading wave packet (another stranger in physics, by the way) localized. And they mean it.

Fame and Authority

This is the main street department of LOP. Everybody should know and accept that there are no "authorities" in science. Experts, maybe. Even the most famous researchers associated with the celebrated sensations of 20th century physics had their forerunners whose names are sadly forgotten, disregarded or even suppressed.

Historical honesty and justified credits do not exactly enjoy the loud voice of the LOP. Examples? *Henry Cavendish* is reported to have discovered the $1/r^2$ force law with his torsion scale decades before *Coulomb.* The bending of a light beam in the gradient of a gravity field was treated by *Georg von Soldner* as early as 1801. In 1898, *Paul Gerber* arrived at the formula on the mercury perihelion precession that later contributed to *Einstein's fame.* Biased criticism on predecessors' work (as e. g. in *Gerber*'s case) is a kind of LOP to avoid priority discussions. $E = mc^2$ has become the *alter ego* of *Einstein.* But there have been researchers before him who dealt with the idea of combining energy with mass and c^2. Our late NPA member *Peter Graneau* found its traces way back in *Wilhelm Weber*'s scriptures. *Paul Wesley* extended *Weber*'s theory to cosmology and arrived at the

"cosmological constant" c^2 thus giving it physical meaning.

For historical justice, *Willy Wien* should be credited for essential preparatory work on radiation that eventually led to quantum physics. *Wien*'s formula $C_1\lambda^{-5}\exp(-C_2\lambda/T)$ has a prefactor $C_1 \sim 5.955*10^{-17}$ Wm² which is the later famous *Planck* quantum of action, h, times c^2. In a mathematical trial-and-error effort to reconcile the formula by *Rayleigh-Jeans* with *Wien's, Planck* replaced *Wien*'s denominator $1/\exp(C_2\lambda/T)$ by $1/[\exp(C_2\lambda/T)-1]$.

The LOP creates celebrities and personal cults. Awards are OK if justified and if not just a decision to erect a monument. The 2014 Nobel Prizes deserve applause. But this has not always been so. With the "Olympic Competition" and the growing number of competitors the pressure on scientists has increased. The war-cry "publish or perish" drives rising generations of scientists to their arms, i. e. to a very special form of the LOP in which the *number* of publications counts (counting in its very original meaning!).

"Εὑρηκα!"

"*Heureka* – I have found it"! *Archimedes* is credited with the discovery of specific weight and this quote marks his lucky moment of discovery. Science is full of stories about those particular moments and scientists certainly need them. But *heureka* moments also seduce to stop further questioning. Incessant questioning is a condition *sine qua non* for science. "*War es ein Gott, der diese Zeichen schrieb*?" (Was it a God who wrote these symbols?) This is *Ludwig Boltzmann*'s famous comment on *Maxwell*'s equations. The (justified) joy over the great success of the first unified theory of electromagnetism must not make us believe that *Maxwell*'s equations are flawless. In fact, the derivation of electromagnetic waves is a masterpiece of mathematical manipulation (mathematical LOP!) and leaves us with severe inconsistencies. The field sources, charges and currents, must be neglected *in order* to derive a propagating wave! *Maxwell*'s equations owe their success to the experimental fact that all electromagnetic phenomena are detected due to the *presence of charges* (source and detector) where those fields occur quite naturally, as in *Heinrich Hertz*'s early experiments. When the theory fails to strictly account for long distance communication between source and detector, there is always photon transport to the rescue. *Planck*'s blackbody radiation formula, *de Broglie*'s wavelength-momentum and *Heisenberg*'s uncertainty relation are symbolic names for other famous *heureka* adventures of physics that are generally accepted beyond questioning. *Einstein*'s photoelectric effect is a brilliant example of how an interpretation of an experimental result becomes a dogma, one supported by a Nobel Prize even. *Einstein*'s "one photon in – one electron out" left no other choice than assigning a photon an internal frequency with all the fatal consequences. Here we smell the origin of the "wave-particle dualism".

Ill-Defined Concepts and Sloppiness

Strangely, "obvious" concepts like *time* and *space* cause considerable problems if not precisely defined. *Time,* for instance, comes in three different meanings: As *co-ordinate* in an arbitrary system of reference ("what time is it?"), as *duration* (an interval defined on a time scale), and as an *abstraction*. This is quite trivial (or should be). Clocks compare *durations* by virtue of counting and displaying periodic processes, they do not "measure" abstract time. No matter what may (may!) affect the internal mechanism of a clock and may affect its display, abstract *time* is not affected. Nor is *space*. The dogmas of "time dilation" and "length contraction" originate from a fatal confusion of *time and space* variables (t,r) with those of a *wave* (*frequency v, wavelength λ*) *in* time and space. Mathematically, it does not make a difference what part of the phase of a wave, $\varphi = (\mathbf{kr} - \omega t)$ with $k = 2\pi/\lambda$, changes due to the *Doppler* effect. Math comes up with its transformations to the "rescue" of relativity. Now the solid(!) interferometer is preferred to contract instead of the light(!) wavelengths due to the *Doppler* effect. For physics, the difference between abstract concepts (space and time) and wave parameters is essential and should be clear.

The difference between particles and waves should be clear, too. Waves are coherent patterns in particle *ensembles*. A single particle has neither a "wavelength" nor a "phase".

Difficulties arise with "inertial systems". Here, the usual kinematic concept à la special relativity is confused with the dynamic principle that stands behind the conservation of kinetic energy ("inertia") under otherwise static conditions (i. e. motion in a constant potential). "Inertial systems" understood as "systems which move at linear uniform speed in which the laws of Nature hold" (note the plural, one of relativists' pets) are quite arbitrary and violate energy conservation.

Relativities and simultaneity are ill-defined unless the distinction between *event* and *observer* is taken seriously (*Tom Phipps* calls them event and operational relativities). Confusing them all too easily leads to the dogma that "all observers are equivalent". This is not true. In order to judge the impressions of an observer in motion, a unique reference has to be identified which is the only consistent way to account for all motions involved. Otherwise the pitfall of "special relativity" messes up all further conclusions.

The "ether hassle" suffers from ill-defined procedures on both sides, pros and cons. (We remember many an amusing discussion during earlier NPA meetings, don't we?) It does not make sense to declare the "ether" dead just because of the MM null result, mainly for two reasons: The null result makes just part of all experimental findings and a light carrying medium must not be confused with a reference system. Referring the velocity of light to a unique reference (no matter whether we call it absolute space or something else) is the one consistent way to account for the conditions of all experimental results. We got used, by the way, to this kind of somewhat unspecified reference when we talk about the velocity of sound in "air". Here the reference is certainly none of the molecules swirling around, yet nobody gives a further thought about the "nothingness" or whatever there might be between the molecules about which we know as much as about the "nothingness" in deep space. Fortunately, we don't have to. An acoustic MM experiment (performed by *Norbert Feist*) gives just the same answer as the MM experiment on light. Clearly, the very existence of physical interactions in both cases cannot be modeled with "nothingness". *Faute de mieux*, we may stick to "field" or "potential" or the "microwave background radiation". That should please pros and cons for the time being. Even *Paul Wesley*, who certainly was not exactly an advocate for any of the ether models, put it this way "It would seem that <u>something</u> must be locally present that causes the locally derived laws of physics to depend on absolute space". Nature provides us with one global system and this is the stage on which Her laws are effective, exclusively and uniquely. Relying on uniqueness, we have a solid basis that does not have to call for additional assumptions.

The LOP beats around the "ether problem" and speaks in several foreign languages, foreign to physics, that is. None of the cult theories is sacrificed and we are left with the peaceful non-scientific coexistence of "no ether", "vacuum fluctuations", "space curvature", and more.

It is customary to speak of "energy quanta", which is not what the minimum product $h \geq \Delta E \ \Delta \tau$ of energy transformed ΔE and the duration $\Delta \tau$ (not "time", please!) of that process says. If at all, it is action that is quantized. But we must be careful here extrapolating our conclusions to a single particle. A lonesome particle cannot change its state all by itself without interactions that make it part of a *complex dynamic system*. This encourages us to ask the provocative question "is quantization an ensemble phenomenon?" Provocative questions are indeed better than permanent propaganda against better judgment.

Moreover, the different faces of h have not yet been explored critically. Is h action, angular momentum, or a criterion limiting the observability of interference? Being quite unspecific, the product "energy×time" defines just the *dimension* of action. It does not have any other physical meaning. Neither does the product "momentum×position". Non-commutativity of certain mathematical tools ("operators") do not give us physical insight. *Heisenberg's heureka* at the crossroads of math and physics entered non-commutativity from math and the quantum h from physics, and math got the right of way. Mathematically, one may juggle around with the "conjugate" parameter pairs (p, r) or (E,t) or with the third choice, (φ, n), phase and number of particles. Introducing *Planck*'s h to the latter may endow it with the dimension of action; this carries "uncertainty" to a particle *ensemble*. Speaking of ensembles:

Wavelength, frequency and phase are *wave* parameters. If we use the definition of the *phase* velocity $c = \nu \lambda$ in a consistent way, we must stay in the picture of a wave as a *particle ensemble* phenomenon. Now the frequency is a *rate of arrival* of photons in a coherent bunch, not the intrinsic frequency of a single photon. Otherwise, the two representations of the black body radiation (in terms of frequency ν or wavelength λ) would not be equivalent.

And we have the case of misleading approximations that look so good that they are never doubted.

A nice classical textbook example showing that math does not care all too seriously about physics is the oblique throw. We don't even have to calculate to see that the path cannot be a parabola, no matter how precise the numerical description. Considering energy conservation suffices to see that the thrown object does not follow an open path when it returns. Again: *A correct result is not sufficient to prove a physical theory right.*

Unfortunately, lacking precision often helps to establish a correct result, thus blurring the mistakes that led to the result. We have to be careful: Particles are not waves, time is not duration, action is not energy, causality is not determinism, active causality is not passive causality, ensemble parameters do not apply to a single particle, etc. If our concepts are not well-defined, physics becomes a playground for fantasy, especially when we have to deal with:

Gedanken Experiments

They are a mighty tool of LOP. *Gedanken experiment* is an oxymoron because it replaces experience by assumptions and conjectures. It is the realm of pseudo answers by virtual reality. Yet gedanken experiments may be very instructive if they stick to the rules of physics. This goes for experiments that comply with a real physical principle but cannot be performed for practical reasons, like an object moving through a hypothetical tunnel through earth's center under the influence of gravitation. But the misuse of gedanken experiments eventually gives a carelessly skeletonized scenario the status of a physical principle. We may have serious doubts about *Einstein*'s "Riding on a light beam" or *Heisenberg*'s "Single photon microscope" or *Schrödinger*'s Cat, and the conjectures thereof.

Simultaneous measurement of position and momentum? In the case of a "free" particle, i.e. in a region of constant potential, its position and momentum have no physical meaning – they are *assumed* to please motivated reasoning which in this case is the non-commutativity of momentum and position operators (LOP of math at work!) and to justify "uncertainty". Where does the particle have its velocity and position from? Why do we have to measure both simultaneously? Trivially, linear momentum changes the position of the particle. But this is no justification to establish an intrinsic uncertainty. Action is a *dynamic* principle which puts the parameter products ($\Delta E \, \Delta\tau$) and ($\Delta \mathbf{p} \, \Delta \mathbf{r}$) on an equal footing, with Δ symbolizing local changes of energy, momentumm position, respectively, and $\Delta\tau$ the *duration* of the process. Here, the operator language gets into trouble – what is a "time operator" supposed to do? The definition of action $\Delta E \, \Delta\tau = \Delta \mathbf{p} \, \Delta \mathbf{r}$ yields a straightforward derivation of the (*neomechanical*, not "special relativistic!) factor $\gamma = (1 - v^2/c^2)^{-1/2}$ with v referring to the unique system of reference characterized by the c^2 potential, no transformation involved. The existence of a unique limiting speed, c, now is the collective effect of the large dynamic system we rightfully call "Universe". A single mass need not diverge with its v approaching c. More likely its dynamic surroundings change, thus associating γ with c^2 rather than with m (mathematically, nothing changes in the formula). A similar argument applies to tunneling. Who (except the LOP) says that a confining potential stays unchanged under the attack of an escaping particle and who says its penetration then must be a "quantum mechanical phenomenon"?

Experiments have never to do with single non-interacting particles. For a gedanken experiment, there seems to be no limit as to the simplification of a chosen scenario. No wonder, this often leads to fruitless discussions.

Fancy Names

Here is a random selection of fancy names that come as trademarks in support of the theories they claim to represent in a nutshell:

Time Dilation; Inertial Systems; Space Curvature; Lorentz Contraction; Superposition; Probability Waves; Self Interference; Collapse of Wave Packets; God Particle; Entangled States; Vacuum Fluctuations; Teleportation; Self Energy; Uncertainty; Dualism; Complementarity; Virtual Particles; and, of course, the notorious "Big Bang" (an originally ironic remark by *Fred Hoyle* who coined it). Some of them are contradictory, some ill-defined, and some are quite funny. Fancy names may originate from a "*heureka!*" experience, but this does not entitle them to block further inspection. For instance, what new insight did *Einstein*'s curvature of space give us that goes beyond *Newton*'s time-honored gravity potential? It is self-contradictory to assume empty space (a "void" or whatever it may be called) and yet equip it with some physical quality if needed.

Misrepresentations

Textbooks, intentionally or not, often confront their readers with a presentation that does not stand the test of scientific scrutiny. Let us take a look at just two classical examples.

Ole Roemer had an ingenious cosmic method to estimate the finite one-way value of c from observations of the eclipse of Jupiter's moon Io under conditions half a year apart, when earth was approaching or receding from Jupiter. Some books on relativity seem to feel uneasy with that "dynamic scenario" (fastest changes of distance in opposite directions) and prefer to make an impression that *Roemer* measured c from distances in opposition ("static scenario" with smallest and greatest distance), which, of course, is impossible with the sun in between. *Roemer* calculated the duration for light to cross the diameter of earth's orbit from his experimental estimate.

The textbook treatment of the *Doppler* effect starts with frequency changes due to (relative) motion between observer and source. This blurs (or even hides) the physics behind the effect which has two primary and quite different scenarios: Observer in motion registers an apparent change of relative velocity and hence an apparent change of frequency; source in motion causes a real change of wavelength because the limiting propagation velocity does not allow the wave to "outrun" the source (*Mach* cones are known for sound and light, aka the *Čerenkov* effect in the latter case). The frequency change is a *secondary* phenomenon in the two cases. The observer in motion is not as privileged as the (usually hypothetical) one at rest. Misused to support the ambiguous "relativity principle", the *Doppler* effect eventually led to the dictatorship of transformations in physics.

Danger – Dogma Ahead!

In spite of its many errors, we do (and should) not imply that the LOP always comes up with wrong answers. The answers may be very well right, but not necessarily the method that led to them. This makes its identification sometimes difficult. The success of some prominent theories is owed to benevolent math. The LOP often defends a correct result, but becomes fatal if it supports contradictions and it becomes dangerous when it turns into a dogma. The Fatal LOP is a FLOP. In the scientists' struggle for fame and honor, the LOP forms a mighty club (in its 2 different meanings!). Excuse the puns – a dogma in science calls for fighting also with the weapon of irony.

LOP Beyond Physics - its Most Rewarding Lesson

The aim of our little excursion through the regime of the LOP is to sharpen our view to its manifold disguises, not just in science. The LOP is a language that claims to advertise the only correct way but fails to encourage objective criticism. The LOP is also the language of importance, pride, and exaggeration. Its dogmatic use is always anti-scientific. Science must not become a battlefield of emotions.

An exhaustive treatment of the LOP in physics, let alone in our everyday lives, is impossible. LOP is a wide field for all kinds of make-believe.

The study of the LOP is a rewarding lesson beyond physics. Our everyday lives are constantly influenced by manipulations from outside or inside (yes, sometimes we apply LOP on ourselves). Next time you go to a supermarket, watch for the signs of the LOP there. You will detect a surprising variety of LOP that lures around many corners. We may learn a lot about the human nature, i.e. about ourselves. We have to accept the LOP where it seems to be unavoidable (just think of its most »valued« kind, money), but we have to avoid it where it certainly does not belong. Where they are least expected, in science, we need whistle blowers. Their objective is not to attack people which is a deplorable emotional aspect of the LOP when it has run out of scientific arguments. Instead, they must point their fingers at objectionable concepts and their fatal consequences - in respect of the one and only authority: *Nature*!

References?

No references will be given. We don't want to be selective here. This essay is just a pathfinder through the jungle of science under the influence of various kinds of LOP. In the time of internet and "globalized wisdom", it suffices to enter a name for more information. But – beware- the virtual digital world is the place of the most widespread kind of LOP ever! It is mandatory to make clever use of the internet as a directory only, not as the source of exclusive truths.

Acknowledgement

I am indebted to *Isia Khalfi* for bringing the lawyers' view of LOP to my attention.

EARTH EXPANSION IS COSMOLOGICAL EXPANSION

VOLKMAR MÜLLER

Retired

08428 Langenbernsdorf, Germany

Standard cosmology sets a lower limit for cosmological expansion. This limit is questioned for reasons both theoretical and observable. The cosmological redshift is not interpreted as a numerical increase of distance but as an expansion of the spatial-temporal scale. Earth and other small objects participate in the cosmological expansion in contradiction to standard theory. This expansion eliminates the need for dark matter to explain the orbital speeds in galaxies. All gravitationally dominated objects, including Earth, take part in cosmological expansion.

Key words: Cosmological Expansion - Earth expansion - Common expansion rate - Examples – conclusions

1. Introduction

How should cosmological expansion be interpreted? Standard cosmology assumes that space itself is expanding and there is no radial-velocity or relative-speed. According to the cosmological principle and the Copernican principle, the observer as well as the observed object (disregarding peculiar speed) can be considered quiescent. The interpretation of the cosmological redshift as relative-velocity is excluded. (The redshift of a white dwarf cannot be used to determine its relative-velocity.) Wikipedia [1] formulates the expansion of space as follows:

The metric expansion of space is the increase of the distance between two distant parts of the universe with time. It is an intrinsic expansion whereby the scale of space itself changes. This is different from other examples of expansions and explosions in that, as far as observations can ascertain, it is a property of the entirety of the universe rather than a phenomenon that can be contained and observed from the outside.

Cosmological expansion is characterized by the Hubble-parameter, 72 + 8 km / s per Mpc (1 Mpc = $3{,}1*10^{19}$ km) [2]. This will not be interpreted to mean that an object will increase its relative speed by 72 km / s for every Mpc that it is away. The numerical distance stays precisely the same. It is the gravitationally dominated space and its measurement that expand. Only objects and areas, that are gravitationally dominated, expand with the space and its measurement. Rationalizing units:

$$\alpha = 72\,\mathrm{km}*s^{-1} / 3{,}087*10^{19}\ \mathrm{km} \cong 2.3*10^{-18}\ s^{-1} \quad (1)$$

Every gravitationally dominated object expands around $2.3*10^{-18}$ per second. The temporal units also grow if the speed of light is constant. After each second, the distances are greater than previous by a fraction of $2.3*10^{-18}$. After $31.56*106$ seconds, or one year, the Earth's radius expands by $31.56*10^6 s * 2.3 * 10^{-18} s^{-1} = 7.26*10^{-11}$. This fraction of the Earth's radius corresponds to about 0.5 mm. But because the expansion is cosmological there is no relative speed between the earth center and its surface. The Earth's radius is therefore numerical constant. The same applies for the distance to the moon. This expands in a certain period of time by a certain fraction. If the time period is, for example, 1.37 billion years: $1.37 * 10^9 * 31.56 * 10^6 s * 2.3*10^{-18}\ s^{-1} = 0{,}0994$. During 10% of the world's age the moon moved away about 10%. (See fig. 1). Analogously the size of the expansion of a gravitational-dominated object can be calculated for every distance and for every time.

When using the analogy of "a balloon" keep in mind: The 2-dimensional surface represents the 3-dimensional volume of the universe. Measurements take place in the universe as on the surface of the balloon. The scale is pasted on the balloon and grows with the balloon. It becomes clear: A measurable, numerical relatively-velocity between two points of the expanding balloon does not exist!

The cosmological expansion is proportional to distance. In addition there may be a positive or negative relative-speed. The relative-speed (peculiar speed) becomes identifiable in the redshift on scales <10 Mpc.

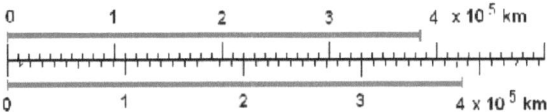

Fig. 1: Measurable distance of the Moon at different times as a one-dimensional example of the expansion of space. If the lower scale corresponds to the recent distance so the upper scale applies to time before 1.37×10^9 years (90% of the world age).

2. The low limit of the cosmological expansion

A.Einstein and E.G.Straus [3] postulated a minimum distance at which the cosmological expansion appears in the Friedmann-Universe. Existing objects within this distance should supposedly show no cosmological expansion. The lower border is determined by comparing (difference-formation) the gravitational potential of a mass with the potential of cosmological expansion. The potential of a gravitating body decreases with the square of the distance. However, the potential of the cosmological expansion increases linearly with the distance. Therefore there is a distance at which both potentials are of the same size. This is the smallest distance where a cosmological expansion is possible in the Friedmann-universe. Systems bounded by gravity are always smaller than this minimal distance. Galaxy clusters, galaxies, stars, planet systems and planets are bounded by gravity and do not participate in cosmological expansion.

Using the earth as an example: The gravity of the earth generates an acceleration of approx. 9.81 m / s^2 at the surface. At a height or distance of approx. 760 * 10^6 km (~ 5.1 AU) the gravitational acceleration amounts to only 6.9 * 10^{-8} cm /s^2. The acceleration by the gravitational field of the Earth at this distance is comparable to the acceleration of the hubble-effect. (Neglecting the potentials of Sun and Milky way). At this distance both potentials are the same size. The standard theory concludes: A test particle at this distance does not fall to the Earth and does not drift away with the cosmological expansion. A test particle is perceived as dormant (Ro = zero-velocity surface). According to standard cosmology, the low limit of the cosmological expansion is reached. Nevertheless, this argumentation of the standard theory has theoretical and observable defects:

The gravitational effect can also be found at distances >Ro. The phenomenon Virgo infall may be mentioned here [18]. One finds in distances < Ro the phenomenon of cosmological expansion. The examples in Section 3 and Table 1 are observable defects in the standard theory in determining the lower limit of the cosmological expansion. While the closeness of any two of these phenomena may be attributed to coincidence, the probability that this many phenomena would agree so closely is near zero.

The low limit of the cosmological expansion is made by comparing two different phenomena. A falling test particle has a relative-speed to the earth. It concerns the speed of moving matter. The cosmological expansion is not moving matter in the sense of SRT, but expansion of space and its units. Gravitational-dominated objects and distance scale expand with the same rate.

The expansion of the space takes place also in shorter distances as Ro. However, the fall velocity is predominant in comparison with the rate of expansion. At Earth's surface and orbit there is no fall velocity or relative velocity of a test particle with respect to the Earth's core. There remains only the cosmological expansion of space. Wu X et al. [4] measure no expansion of the Earth because they are measuring with units that expand with the same rate. The radius of the Earth behaves as the spatial structure requires it.

Only things which are not dominated by gravity do not expand with the space. These contract in relation to gravitationaldominated objects and distances. Continents contract in relation to the size of the Earth with the expansion rate of the universe. The limiting magnitude of Ro is, according to [18] and others exceeded by gravity. Ro is according to Section 3 exceeded by cosmological expansion in the opposite direction. The distance Ro is obvious for both phenomena no limit. The low limit of the cosmological expansion should not be where the gravitational and expansion potentials are equal but where the electromagnetic forces play a greater role than gravitational.

3. Examples of cosmological expansion on small scales

3.1. *Expansion of early galaxies*

Galaxies at a distance of 10.7 billion light years (z ~ 2.3) were examined by P.van Dokkum et al [5] and found to be much denser than galaxies in the present (near) universe. The simplest explanation is that all galaxies were once that dense and that they expanded until they arrived at their present density. Calculations by V.Müller [6] give a necessary expansion rate of $2.37*10\text{-}18s^{-1}$ in agreement with cosmological expansion. Investigations by P.Oesch et al. [7] provide similar expansion values for still more distant galaxies and support this assumption. A cosmological expansion of these galaxies lowers the border postulated by the standard theory.

3.2. *Rotation of galaxies and clusters of galaxies*

The orbital velocities of objects in galaxies remain fairly constant with increasing distance from the center despite Kepler's 3rd law (Y.Sofue and V.Rubin [8]). The standard theory explains this curve by postulating a high density of "dark matter" in the outer areas of the galaxies. Some authors prefer a modification of Newton's dynamics (e.g. MOND) that does not require dark matter. Neither procedure is used here. We accept the following requirements:

1. There is a cosmological expansion in small areas.

2. The cosmological principle is maintained. Observers and objects in the universe may be considered dormant (Peculiar speed may be neglected). The cosmological expansion has no relative velocity in the sense of the SRT.

According to requirement 1) the orbital paths go away from the galactic centers as a result of cosmological expansion. According to 2: Between the galactic center and the object in orbit, the space expands. The numerical distance does not! A relative or radial-velocity is not given, however, the spatial units of the abscissa in Figure 2 expand and drift to the right. With numerically constant distance the orbital speed remains constant. The orbital speed does not drop during the expansion because the expansion is not a numerical change of distance. The apparently observed contradiction to Kepler does not exist. Dark matter is not necessary.

3.3. *Developmental effects in the early universe*

In the study of distant supernovae (SN Ia) by Riess A. et al.[20] and Perlmutter, S. et al.[21] noted that there is in these seemingly lower absolute magnitude than at

nearby events of this type. This is very unlikely. In validity of the distance module, the distance should therefore be greater than is given by the Hubble parameter. This had to be therefore smaller earlier. This leads to accelerated cosmological expansion and dark energy.

The author assumes the following scenario: A SN Ia observed at a distance z = 2.3 ($\cong 10.7*10^9$ Ly). The emission was made after about 20% of the current age of the World. In gravity-dominated area, the first spatial and temporal unit is after the emission 20% of the current size. Substituting

$$\Sigma = 0.5n\ (\ x_1 + x_n) \qquad (2)$$

(Σ = Sum of all units, n = $10.7*10^9$ = number of primary units, x_1 = 1 = Scale value of the first unit after emission, x_n=5=Scale value of unity in the observation)

we obtain a light travel time (look back time) of 32.1 * 10^9years. This is the 3 fold value of $10.7*10^9$ Ly. Because the distance is significantly larger, and the apparent magnitude (m) is constant, should be variable in the distance modulus, the absolute magnitude (M). In contrast, the author assumes that the factor (m - M) is constant and the scale value of the distance is variable (x_1 = 1; x_n= 5). Because of the variable scale value of the time scale a constant Hubble number is possible and dark energy is not required.

It should be noted: The radius of the galaxies in Section 3.1. appears as it can be calculated by extrapolation of the UT scale. This does not mean the numerical radius (or distance) has changed but the spatial and temporal units are expanded. The numerical distance remain 32.1 * 10^9Ly. This corresponds to 10.7 * 10^9 Ly current size. The same is true for example for the Earth: After 2/3 of the world ages (4.2 *10^9 years ago) was the Earth's radius 2/3 of the todays Earth radius (section 3.7.). But measurements in the UT-scale yield constant radius of the Earth. Likewise, the galaxies in Section 3.1. or other distances.

The number density of quasars should be by P. Schneider [22] at z ~ 2-4 more than 100 times greater than today. At approximately the same z-value is obtained approximately the same density values as described in Section 3.1. . This points to the same cause. One sees as cause obviously a rapid development in the early universe. Here cosmological expansion in the sense of expansion of units is assumed analogous to Section 3.

In applying the above summation formula (2) there are no problems with the formation time of old metal-rich objects in the early universe. This also applies to oldest globular clusters.

With respect to the density of starbursts at distances> z ~ 1 large values are similar, have been observed. There is extensive literature eg [23], [24]. The radii of the studied galaxies appear to behave inversely proportional to increasing z - and density values. If there UT - scale applicable, so there was no change in distance between the objects and no abnormal density values. Within the galaxies, ie gravitationally bound systems, the cosmological expansion accordingly the adoption of the standard cosmology should not be existent. If dark matter exists, then the gravitationally bound radii are several times larger. Can be there is really still areas in which cosmological expansion occur?

3.4. *The pioneer anomaly*

The pioneer 10 and pioneer 11 probes were launched in 1972 and 1973. Both probes experienced an anomalous negative acceleration directed toward the center of our solar system. This acceleration was calculated by J.D.Anderson et al [9] to be $8.74 \pm 1.33 * 10^{-8} cm/s^2$. Similar anomalies were found for the Ulysses and Voyager probes.

If the anomalous acceleration is divided by the speed of light we get a value close to the Hubble-parameter:

$\alpha = 8,74*10^{-8}cm*s^{-2}/299,792*10^8 cm*s^{-1} = 2,92*10^{-18}s^{-1}$

A phenomenon of the same order of magnitude as the cosmological expansion affects space probes at distance scales of the solar system. However, the phenomenon manifests itself as blue shift of the radio signals. If this effect is due solely to cosmological expansion, a redshift would be expected. Why this is so remains unclear for the time being. The pioneer anomaly and Hubble constant agree to 18 orders of magnitude.

The author speculates that the space and units also expand in the solar system. In the past the spatial dimensions were smaller. Because of the steady speed of light, the temporal measure was also smaller (second was shorter). A spacecraft that started ~40 years ago would have shorter seconds and shorter meters. This

would make the numerical whole distance seem shorter than with steady spatial units. This corresponds to a deceleration; exactly what was observed with the Pioneer probes 10 and 11.

3.5. *Mars rotation*

It is generally assumed that the earth's rotation is slowed down by tidal friction because of its large moon. If the tidal friction delays the rotation of the Earth, this delay is earth-specific. Mars has only two very small moonlets. The tidal friction is negligible. But if the rotation delay of the earth is caused by cosmological expansion, this also applies to Mars. If both planets show the same secular delay rate then the likely cause would be cosmological. According to unconfirmed statements now common change in day length of Earth and Mars was found [19].

3.6. *Expansion of the lunar orbit*

O.J. Dickey et al. [10] using LLR technology measured the present expansion of the lunar orbit to be approximately 3.82 ± 0.07 cm / year. This gives an expansion rate for the lunar orbit:

$$\alpha = \Delta r *(r * t)^{-1} \qquad (3)$$

$\alpha = 3, 82$ cm $(3,844 *10^{10}$ cm $* 31,56 *10^6 s)^{-1} = 3,148 \pm 0,058 *10^{-18}s^{-1}$

α =expansion rate, Δr = expansion factor (3.82 cm), r = distance (3,844 *10^{10} cm), t=period (1a =31.56 *10^6 s).

The expansion rate of the lunar orbit is approximately the expansion rate of the universe. It is slightly higher most likely because of tidal friction and other gravitational interactions with the Earth and Sun.

3.7. *Cosmological expansion of the earth radius*

A Mpc amounts to approximately $3*10^{19}$ km. The earth radius amounts to approximately $6*10^3$ km. The earth radius is $5*10^{15}$ times smaller than 1 Mpc. Assuming that the cosmological expansion speed is proportionally to the distance, the earth radius expansion would be $5*10^{15}$ times smaller than 72 km/s. Converting units we get approximately 0.06 cm/year for the expansion of the earth's radius. Measurements of Earth's rotation delay revealed a recent value of 0.0016 s / cy (N.Bär [11]). This corresponds to a delay rate of $2.93*10^{-18}s^{-1}$. This value is again roughly the expansion rate of the universe (Hubble constant.)(See table1). Rearranging (3) we get

$$\Delta r = \alpha * r * t \qquad (4)$$

(Δr =expansion factor, α =expansion rate $(2.93*10^{-18}s^{-1})$, r = earth radius ($6371*105$ cm), t=period ($1a =31.56 *10^6$ s)

The expansion rate of the Earth becomes 0.06 cm / a.

As the earth expands its moment of inertia increases and if angular momentum is conserved, its rotation rate should decrease. Therefore, if we know the rate at which the rotation slows, we can calculate the rate the radius expands. Using the twist theorem (pirouettes effect):

$$\Delta r = r \ [(1 + t / t)^{0.5} -1] \qquad (5)$$

(Δr = radius difference, r = earth radius (cm), t = period (s), Δt = time difference ET-UT)

$\Delta r = 6371 * 10^5 \ [(1+ 0.0016 / 86400)^{0.5} -1 \ ] = 5.9$ cm/cy = 0.059 cm / a.

Using the Pirouette effect, the rotation delay of the earth over 100 years corresponds to an expansion of the earth radius of 5.9 cm. (The value t = 0.0016 / 86400 refers to the lengthening of the day per 100 years.) This value is equal to the calculated value of Earth expansion by L.Egyed, H.G.Owen [12], V.Müller [13]. The cosmological expansion, 72 km $* s^{-1} * $Mpc^{-1}, corresponds to 0.05-0.06 cm $* a^{-1} * r_E^{-1}$ (r_E = radius of the Earth). The author therefore assumes a cosmological origin to earth expansion. Wu et al [4] tried to determine the rate of earth expansion using the International Terrestrial Reference Frame which incorporates satellite laser ranging and the Very Long Baseline Interferometry. They limited the rate of expansion to be less than 0.2 mm/year. This is consistent with the above analysis. However, they found a mean drift of their calculated reference frame relative to the earth center of mass of 0.6 mm/year. That is, the center of their reference frame had to be continuously corrected at the rate of 0.6 mm/yr. The origin of this drift is not obvious. If we calculate this drift as a fraction of the Earth's radius we get: 0.06cm/($31.56*10^6$s$*6371*10^5$ cm) = $2.98 * 10^{-18}$s-1, which is close to the Hubble parameter so we might assume that its origin is cosmological.

3.8. Area ratio oceans / continents

Today's continental surfaces, including shelves, amount to approximately $177*10^5$km^2. According to Hilgenberg, Carey, Vogel et al, this surface can fit on a globe of approximately 3750 km radius. They will completely cover this globe without significant overlaps or empty spaces. This can be best explained by an expansion of an earlier, smaller Earth. The continental surface could have been covered by water. If in relation (3), one inserts 6371 km for the present earth radius, 2621 km for Δr (6371 - 3750 = 2621 km); and $4.3 *10^9$ years for the age of the continents; one gets an expansion rate of $\alpha = 3.0 * 10^{-18}s^{-1}$. This approach results in a cosmological expansion of the Earth, but not the continental areas. This is because the Earth is dominated by gravity, while the continents are dominated by electromagnetic forces and do not expand with the cosmos. How the earth complies with cosmological expansion is interesting, but not relevant (ocean floor spreading, subduction etc).

3.9. The inner core of the Earth

The hypothesis of S.K.Runcorn [14] led to the formation of an inner core with a radius growing at 121.5 km/Ga or 0.01215 cm/year. The "surface" of the earth's inner core "goes away" from the earth's center at approximately 0.01215 cm per year or $3.85*10^{-10}$cm /s. One divides this value by the earth-core radius and receives the expansion rate of $3.85*10^{-10}$ cm$*s^{-1}$ / $1270*10^5$cm = $3.0*10^{-18}$ s^{-1}which is very close to cosmological expansion. The expansion rate of the inner earth's core and of the universe are roughly the same. The value theoretically investigated through S.K. Runcorn may not correspond to all modern opinions. It is noticed however that C.Denis et al [15] receives similar results and supports Runcorn's presumption.

4. Conclusion

The redshift in spectra of extra-galactic objects is proportional to distance. It is not a radial velocity in terms of the SRT. Numerical distance does not grow with cosmological redshift, but the spatial units do grow. Earth's core, Earth, lunar orbit and galaxies do not expand in the space but with the space. The standard cosmology sets a lower limit to this expansion. Meanwhile, there are several authors who notice the coincidence of the found values with the Hubble constant [25]. Preceding examples show that this limit is where gravity loses dominance to electromagnetism.

Earth, larger areas, and the speed of light obey the structure of the space. In the SI system everything expands that is dominated by the spatial structure (gravity). Earth's crust, Planetoids ($<\sim$ 200 km diameter) and everyday objects are dominated by electromagnetism and do not expand in the SI system.

One can define space time measures on an electromagnetic basis (SI second) or on the basis of gravity (UT second). The SI second and on it based dimensions (length, speed) are not dominated by gravity and do not expand. Both scales do not run synchronically.

The divergence amounts $\sim 2.7(\pm\ 0.4\ )* 10^{-18}\ s^{-1}$. With use one receives different measuring results. The standard theory uses the SI system for gravitationally dominated areas as well as for electromagnetically dominated areas.

In summary, when using the UT second to describe gravity dominated objects one observes:

- These objects have numerically constant

rotation   periods, radii and densities in the UT-scale.

- The Earth is not expanding into space, but space expands with the Earth.

- Cosmological radial velocities do not correspond with the SRT and are cancelled.

- Dark energy is not necessary to describe the magnitude of Supernovae.

- Dark matter is not necessary to describe the rotation of galaxies. Modified Newtonian dynamics is also not required

- There is no cosmological numerical expansion.

- The universe is numerically unlimited in time and space.

The above examples are suitable to establish the above conclusions.   Last but not least, the hypothesis of .W.Carey is supported by these conclusions: Continents do not expand with the whole Earth which does expand. He hypothesized that the origins of earth expansion were cosmological. He might be right.

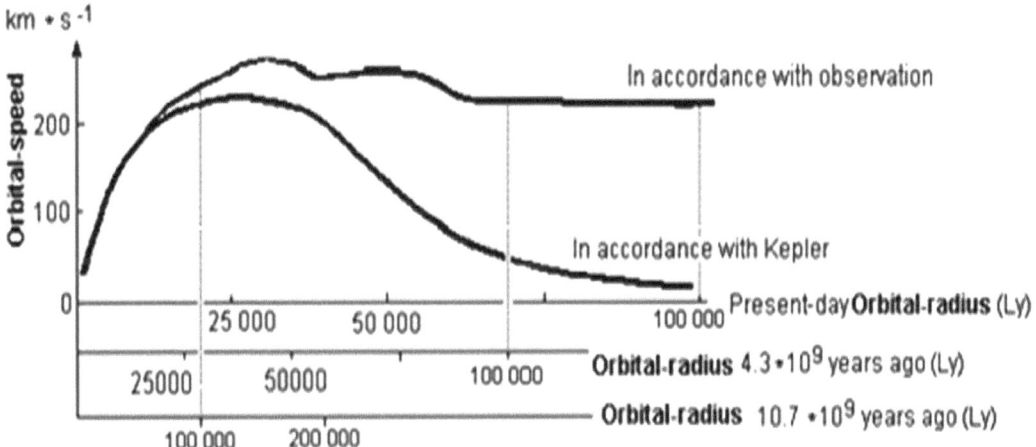

Fig 2: The values on the abscissa are subject to adrift of the size of the cosmological expansion rate. An object in 30 kpc (~100 000 Ly) distance from the center has today almost the same orbital speed as before 4.3 * 109 a or 10.7 * 109 a because the numerical orbital radius will not change (interference forces neglected). Keplers law is also not injured without dark matter because the numerical orbital radius is constant.

Table 1. **Examples of the occurrence of the cosmological expansion rate (Hubble constant) in the forbidden area**

1	2	3	4
Example	**Rate**	**Possible other causes**	**References ***
Expansion-rate of the universe	2.33 ± 0.26 x 10-18s-1	-	[2]
Expansion of early galaxies	2.37x 10 -18s-1	?	[5]
Pioneer anomaly	2.91 ± 0.44 x 10-18s-1	Thermal radiation pressure	[9]
Expansion of the Moon orbit	3.15 ± 0.06 x 10-18s-1	Tidal friction	[10]
Delay of the earth's rotation	2.93x10-18 s-1	Tidal friction	[11]
Surface relation oceans/continents	3.0 x 10-18s-1	?	[6]
5-dimensional field theory	3.6 x 10-18s-1	Scale effect	[16]
Polar diameter of Earth	$2.5 \pm 0,95$ x 10-18s-1	Post glacial uplift	[17]
Inner earth's core	3.0 x 10-18s-1	Growth by phase-conversion	[14]

* In column 2 contained values is calculated by the author. Calculation bases are informations out of column 4.

Acknowledgments

I thank Martin Kokus for preparatory discussion and help with translation.

References

1. Wikipedia [Online] Metric expansion of space http://en.wikipedia.org/wiki/Metric_expansion_of_space (accessed Feb.12, 2014).
2. W.L. Freedman, B.F.Madore, B.Gibson, L.Ferrarese, D.D.Kelson, S.Sakai, J.R.Mould, R.C.Jr.Kennicutt, H.C. Ford, J.Graham J.Huchra, S.Hughes, G.D.Illingworth, L.Macri, P.B.Stetson, *Astrophysical Journal,* **Band** 553, 47 (2001)
3. A.Einstein.and E.G.Straus, *Rev. Mod. Phys.* **17**, 120-124 (1945).
4. Wu, X.; X.Collilieux, Z. Altamimi, B.L.Vermeersen, R.S.Gross, I.Fukumori, Accuracy of the International Terrestrial Reference Frame origin and Earth" *Geophysical Research Letters*, **38**, 5 (2011) L13304.
5. P.v Dokkum, M.Franx, M.Kriek M, B.Holden, G.Illingworth, D.Magee, R.Bouwens, D.Marchesini, R.Quadri,G.Rudnick, E.Taylor and S.Toft. (2008), Confirmation of the remarkable compactness of massive quiescent galaxies at z~2.3". *Astrophysical Journal*, **677**: L5-L8 (2008).
6. V.Müller "The Cosmological Expansion of Small Regions and of the Earth" in : G.Scalera : Selected Contributions to the Interdisciplinary Workshop THE EARTH EXPANSION EVIDENCE , Aracne editrice S.r.l. Roma, 227 (2012)
7. P.A.Oesch; R.J.Bouwens; C.M.Carollo; G.D.Illingworth; M.Trenti; M.Stiavelli; D.Magee;
8. Y.Sofue and V.Rubin „Rotation curves of spiral galaxies"Annu. *Rev.Astron.Astrophys.* **39**, 137 (2001).
9. J.D.Anderson, P.A.Laing, E.L.Lau, A.S.Liu, M.M.Nieto, S.G.Turyshev. [Online], Study of the anomalous acceleration of Pioneer 10 and 11". (Dated: 11.April 2002) http://www.arXiv.org/PS_cache/gr-qc/pdf/0104/0104064v5.pdf (accessed Feb 12, 2014)
10. J.O.Dickey, P.L.Bender,J.E.Faller,X.X.Newhall, R.L.Ricklefs, J.G.Ries, P.J.Shelus,C.Veillet, A.L.Whipple,J.R.Wiant, J.G.Williams, C.F.Yoder „Lunar laser ranging:A continuing legacy of the Apollo program". Science,265, (1994) pp 482
11. N.A.Bär [Online] „Die Akzeleration",(Dated:Apr 11,2013) http://www.nabkal.de/akzel.html (accessed Feb 12,2014)
12. H.G.Owen, „The Earth is expanding and we don't know why", *New Scientist*. **22**, 27 (1983).
13. V.Müller, „Does cosmological expansion exist in smaller scale?" NCGT Newsletter No. 50 (2009),
14. S.K.Runcorn, "Towards a theory of continental drift." *Nature,***193**, 311–314 ."Convection currents in the Earth's mantle." *Nature*, **195**, 1248 (1962)

15. C.Denis, K.R.Rybicki and P.Varga,"Secular changes of LOD associated with a growth of the inner core", *Astron.Nachr.* **327**, No.4, 309 (2006)

16. E.Schmutzer, „ Approximate global treatment of the expansion of the cosmic objects induced by the cosmological expansion", *Astron. Nachr.*, **321**, 227 (2000).

17. H.Ruder, M.Schneider, M.Soffel, „Geodäsie und Physik", Physikalische Blätter Nr **46**,No.2,41 (1990

18. I. D. Karachentsev and O. G. Nasonova,[Online] „The observed infall of galaxies towards the Virgo cluster", (Dated: February 9, 2010) http://mnras.oxfordjournals.org/content/405/2/10 75.full (accessed June 25, 2014)

19. P. Christensen [Online] Ask Dr C. http://beamartian.jpl.nasa.gov/towhnall/question/ 350/is-mars-rotation-slowing-akin-to-earth-s-rotational-slowing-could-such-a-feature-indicate-mars-having-a-molten-core (accessed Jul 5. .2014)

20. Riess A.et al.(1998) „OBSERVATIONAL EVIDENCE FROM SUPERNOVAE FOR AN ACCELERATING UNIVERSE AND A COSMOLOGICAL CONSTANT" AJ 116:1009-1038

21. Perlmutter,S.et al [Online] „MEASUREMENTS OF Ω AND Λ FROM 42 HIGH-REDSHIFT SUPERNOVAE" (accessed Feb 04, 2015) *http://arXiv.org/abs/astro-ph/9812133v1.*(1998)

22. Schneider P.[Online]„Überblick Kosmologie" http://www.astro.uni-bonn.de/~peter/cosmo_short.pdf (accessed Feb 04, 2015)

23. Van der Wel A.,[Online] „Extreme emission line galaxies in candels: Broad-band selected, star-bursting dwarf galaxies at z>1" http://www.spacetelescope.org/static/archives/rel eases/science_papers/heic1117.pdf (accessed Feb 04,2015)

24. Hubblesite News Release Archive, [Online] News ReleaseNummer:STScI-1996-37(VillardR.andMadauP.) http://hubblesite.org/newscenter/archive/releases/ 1996/37/ (accessed Feb 04,2015)

25. Ellman R. [Online] "On Five Independent Phenomena Sharing a Common Cause" http://www.the-igin.org/FivePhenomenaCommonCause.pdf (accessed Feb 04 2015)

THE 'HIDDEN VARIABLES' IN QUANTUM MECHANICS. THE PRECESSION OF THE ELECTRON'S SPIN IN AN ATOM

NINA SOTINA
Ph.D. in Physics
Email: nsotina@gmail.com

The author of the present work is a follower of the causal interpretation of quantum mechanics. The causal interpretation agrees with quantum formalism only if "nonlocal hidden variables" are taken into consideration ("nonlocal hidden variables" can be an unknown yet physical field which allow for interactions to attain speeds greater than the speed of light). In this work a critical analysis of Schrödinger's article: "Quantization as an Eigenvalue Problem" (1926) and Bohm's ideas about "a quantum potential" are given. It is proven from the laws of classical mechanics, that the time-independent Schrödinger equation extracts from solutions of the Hamilton-Jacobi equation (1) only those solutions that satisfy a necessary condition of stability under small perturbation forces, and (2) those for which the orbital angular momentum of a particle moving along a closed trajectory is quantized. It is demonstrated, on the basis of the causal interpretation of quantum mechanics, that an electron's spin in an atom precesses. If one assumes that the "hidden variables" are similar in their properties to superfluid, then, in accordance with the properties of superfluid, the electron's motion in an atom must form structures (quasi-particles) in the physical vacuum. It is shown that in this case the natural frequencies of the atom are the frequencies of precession of the quasi-particle's spin.

1. Interpretation crisis of quantum mechanics

Currently, there exists an interpretation crisis in quantum mechanics that is, in essence, due to the lack of comprehension of the underlying physics hidden behind the equations. While the mathematical formalism of quantum mechanics describes many experiments well there are heated discussions among scientists about its physical interpretation. The various interpretations offer different approaches to the issues that arise, which include *the wave function collapse, paradoxes such as EPR*, and etc. However, at present time there is no physical interpretation of quantum formalism that would not have contradictions within it or with accepted ideas and theories. Such situation in quantum mechanics is an important argument in support of the necessity of essential changes in physics; perhaps, the change of the scientific paradigm.

To support our reasoning let us assess two interpretations: the Copenhagen and the causal interpretations. Followers of **the Copenhagen interpretation** insist on the point of view that physics is the science which rests solely with measurements. Thus, under this interpretation it is invalid to making any statements about the systems' properties prior to measurement. In the Copenhagen interpretation joint probability of noncommuting operators cannot be used because direct measurement experiments cannot be conducted. However, followers of the Copenhagen interpretation have difficulty explaining "the essentially quantum effects" (e.g., teleportation of polarization of the photon). In the context of these experiments the problem of interpretation of the quantum formalism is further aggravated: it is necessary to assume that, **although some properties of reality exist before measurements only potentially** (for example, polarization of each photon from the polarization-correlated photon pairs) **however, there is a correlation between them**. If one assumes that quantum objects have a priori properties corresponding to noncommuting operators, either negative probabilities, or «hidden variables» should be introduced in quantum mechanics.

The causal interpretation of quantum formalism centers about the existence of «hidden variables». Historically, the issue of incompleteness in the description of physical reality by quantum mechanics was put forward for the first time by Einstein, Podolsky, and Rosen in 1935 (*the EPR paradox*). They proposed the existence of «hidden variables», which uniquely characterize the given state of the system; thus, allowing a quantum system to be consistent with the deterministic theory. However, in about 50 years John S. Bell advanced his famous inequalities. It followed from the violation thereof in quantum theory that any theory of «hidden variables» claiming to be able to describe experimental results, must be "nonlocal" only. "Nonlocality" means two possibilities: either existence of a physical field which allows for interactions to attain speeds greater than the speed of light {introduction of such a field in physics will obviously contradict the theory of relativity}, or propagation of "signals" of the changes of a particle's quantum state with an infinite speed, in essence, the possibility of a long-distance forces acting over free space.

In order to "explain" the experiments with "the essentially quantum effects," while at the same time avoiding introducing long-distance forces or a field which allow speeds greater than the speed of light in quantum mechanics, physicists began to talk about non-separability of quantum mechanics (in other words about existence some type of an information "link", between remote quantum object). Don Howard wrote in his book [1] that, quantum mechanics is non-separable, local theory {separability means that spatially separated systems can exist in different states. Locality assumes that the state of the system may only be modified through effects propagating at sublight speeds}. It is mistaken to believe that the ideas of non-separability of quantum mechanics and the existence of long-distance forces do not contradict the theory of relativity. Note that to measure the speed of light it is necessary to have a receiver and a transmitter that are not only separated in space, but are also autonomous in their behavior.

The author of the present work is inclined to the causal interpretation of quantum formalism. Using the causal approach it will be proven in this work that the Schrödinger equation can be derived from the deterministic laws of classical mechanics. Using this approach it is also possible to make some conclusions about properties of the "hidden variables".

D. Bohm was the first who introduced nonlocal hidden variables in quantum mechanics [2]. We will show further that some of the assumptions in the Bohm's work were incorrect.

A work of Von Neumann is often mentioned in this regards in which he proved the impossibility of any «hidden variables» in quantum mechanics. In 1935 Grete Hermann published an article in which she exposed an apparent mistake in the Von Neumann's prove. This article remained unnoticed by much of the scientific community for some time; however the mistake was once again independently verified in 1966 by John S. Bell.

2. The work of Schrödinger "Quantization as an Eigenvalue Problem"

On January 27, 1926, an article under the title "Quantization as an Eigenvalue Problem" was submitted by Erwin Schrödinger to the journal "Annalen der Physik" [3]. In this article the equation describing the quantum levels of energy of a non-relativistic hydrogen atom was presented for the first time.

Schrödinger was inspired by the de Broglie's idea of "matter waves". Originally he thought the "matter waves" to be real and envisioned a particle as an actual wave packet. In essence, Schrödinger was, as de Broglie, a supporter of the causal interpretation of quantum phenomena.

To explain the quantization procedure Schrödinger turned to classical mechanics. He started with the Hamilton-Jacobi equation (in the case when potential energy does not depend on time t explicitly). This equation is valid, when a mechanical system is under action of a conservative force $\overline{F} = -grad\,U$. The usage of the Hamilton-Jacobi equation was not accidental: this equation allows reduction of a classical dynamics problem to a solution of a partial differential equation. Utilizing his equation, Hamilton demonstrated the optical-mechanical analogy in geometrical optics: the trajectory of a particle and the trajectory of the light beam coincide when the value of the potential energy of the particle corresponds to the value of a refractive index of the medium.

Schrödinger saw a deep physical meaning in the optical-mechanical analogy and made an attempt to establish an analogy between some parameters of the particle's motion and parameters of the wave packet.

As the starting point of his study Schrödinger considered the Hamilton-Jacobi equation for the motion of an electron in a hydrogen atom. The Hamilton-Jacobi equation for the motion of an electron with mass m_e in a hydrogen atom has form

$$\frac{1}{2m_e}\left[\left(\frac{\partial S}{\partial x}\right)^2 + \left(\frac{\partial S}{\partial y}\right)^2 + \left(\frac{\partial S}{\partial z}\right)^2\right] - \frac{e^2}{r} = \varepsilon, \qquad (1)$$

where S is the Hamilton's principal function , and e is the charge of an electron. Schrödinger next introduced a new function ψ in order to replace function S according to the following substitution

$$\psi(\overline{r}) = \exp(S/\kappa). \qquad (2)$$

After substituting Eq. (2) into Eq. (1) the latter becomes a quadratic form with respect to ψ and its derivatives. Schrödinger next searched for real, single-valued, bounded function ψ that would give extreme value to the integral of the quadratic form, taken over the entire configuration space

$$\delta\int\left[\left(\frac{\partial\psi}{\partial x}\right)^2+\left(\frac{\partial\psi}{\partial y}\right)^2+\left(\frac{\partial\psi}{\partial z}\right)^2-\frac{2m_e}{\kappa^2}\left(\varepsilon+\frac{e^2}{r}\right)\psi^2\right]d\tau=0. \quad (3)$$

The Euler–Lagrange equation for this variational problem is equation

$$\frac{\kappa^2}{2m_e}\Delta\psi+(\varepsilon+\frac{e^2}{r})\psi=0, \qquad \Delta\text{-laplacian}\quad (4)$$

where ψ satisfies an additional condition, which Schrödinger represented as

$$\int|\psi|^2 d\tau=1. \quad (5)$$

Note, that variational problem (3) in Schrödinger's approach substitutes the Bohr-Sommerfeld quantum rules. Setting $\kappa=h/2\pi$ in Eq. (4) Schrödinger then obtained, the well-known, Bohr's energy

$$\varepsilon_n=-\frac{2\pi^2 e^4 m_e}{h^2 n^2}\quad n=1,2,3,... \quad (6)$$

Note, that Planck's constant h was introduced into the analysis 'artificially' to have an agreement with experiments.

For an arbitrary potential energy U Eq. (4) takes the form

$$(h^2/8\pi^2 m)\Delta\psi(\overline{r})+(\varepsilon-U)\psi(\overline{r})=0 \quad (7)$$

which is currently known as the time-independent Schrödinger equation.

Eq. (7) was accepted by scientific community almost immediately. Schrödinger suggested a clear method of finding the quantum energy levels of an atom. Moreover, in Schrödinger's opinion, his method revealed the nature of quantization. He wrote in his work: "It is quite natural to associate the function ψ with some oscillation process in atom".

However the derivation of the Eq. (4) from the laws of classical mechanics, presented by Schrödinger, seemed unclear and even erroneous to physicists for the following reasons.

a) *Substitution* (2) *assumed that ψ is a real function, while Eq. (4) had complex solutions.*

Note here that for the set of solutions of Eq. (4) to be a subset of solutions of the Hamilton-Jacobi equation with a given potential, the following substitution must have been made instead of substitution (2):

$$\psi(\overline{r})=\exp(iS/\kappa). \quad (8)$$

Indeed if we substitute it into Eq. (4), and separate real and imaginary parts, we obtain two equations: the Hamilton-Jacobi equation (1) and an additional equation $\Delta S=0$. If, however, ψ in the form (8) is substituted into integral $\int|\psi|^2 d\tau$ taken over the entire space, the integral diverges. Thus, there are no solutions of the Hamilton-Jacobi equation (1) in the unbounded space which, under substitution (8), satisfy Schrödinger Eq. (4) with condition (5). Therefore **Kepler's orbits which are solutions of Eq. (1) are not solutions of the Schrödinger Equation (4) with condition (5).**

b) *The set of solutions of Eq. (7) with condition (5) is not a subset of solutions of the Hamilton-Jacobi equation with a given potential U if motion occurs in unbounded space.*

Statement (**b**) will be proven in the subsequent chapter. It follows directly from this statement that the analogy between the motion of a material particle under action of a given potential U and the motion of a wave described by the Schrödinger Equation (7) does not exist.

c) *The physical meaning of function ψ was not clear.*

Note that, while developing the concept of the "matter waves", neither de Broiglie, nor Bohm concretized the material medium in which these waves propagate.

d) *The physical meaning of the variational principle used in the derivation of the equation Eq. (4) was not clear.*

For the reasons listed above the **Schrödinger Equation was taken as a postulate.** Thus, mathematical ties with classical mechanics were broken and a new field of science for description of the microworld phenomena emerged - quantum mechanics. In the same year, 1926, M. Born proposed the probabilistic interpretation of wave function ψ.

Below the derivation of the Schrödinger Equation from the laws of classical mechanics is presented, using the method utilized by Schrödinger in his work [3], but correcting mistakes that he made. In addition the physical meaning of the variational principle used by Schrödinger will be given.

3. From the Schrödinger Equation to Equations of Classical Mechanics

In the present work we limit our study to the time-independent Schrödinger equation. This is done to avoid cumbersome equations and in order to present the main idea in an unambiguous manner. The method presented below is applicable to the time dependent Schrödinger Equation without principle difficulty.

Let us consider now more general substitution

$$\psi(\overline{r}) = A(\overline{r}) \exp(2\pi S i / h) \quad . \qquad (9)$$

Substitution (9) leads to the following equations

$$\frac{1}{A} \exp\left(-2\pi S i / h\right) \frac{\partial^2 \psi}{\partial x^2} = \frac{1}{A} \frac{\partial^2 A}{\partial x^2} -$$
$$-\frac{(2\pi)^2}{h^2}\left(\frac{\partial S}{\partial x}\right)^2 + \frac{2\pi i}{h}\left(\frac{\partial^2 S}{\partial x^2} + \frac{2}{A}\frac{\partial A}{\partial x}\frac{\partial S}{\partial x}\right)$$

$$\frac{1}{A} \exp\left(-2\pi S i / h\right) \frac{\partial^2 \psi}{\partial y^2} = \frac{1}{A} \frac{\partial^2 A}{\partial y^2} -$$
$$-\frac{(2\pi)^2}{h^2}\left(\frac{\partial S}{\partial y}\right)^2 + \frac{2\pi i}{h}\left(\frac{\partial^2 S}{\partial y^2} + \frac{2}{A}\frac{\partial A}{\partial y}\frac{\partial S}{\partial y}\right) \qquad (10)$$

$$\frac{1}{A} \exp\left(-2\pi S i / h\right) \frac{\partial^2 \psi}{\partial z^2} = \frac{1}{A} \frac{\partial^2 A}{\partial z^2} -$$
$$-\frac{(2\pi)^2}{h^2}\left(\frac{\partial S}{\partial z}\right)^2 + \frac{2\pi i}{h}\left(\frac{\partial^2 S}{\partial z^2} + \frac{2}{A}\frac{\partial A}{\partial z}\frac{\partial S}{\partial z}\right)$$

Adding these equations we obtain

$$\frac{\Delta \psi}{\psi} = \frac{\Delta A}{A} - \frac{(2\pi)^2}{h^2}\left[\left(\frac{\partial S}{\partial x}\right)^2 + \left(\frac{\partial S}{\partial y}\right)^2 + \left(\frac{\partial S}{\partial z}\right)^2\right] +$$
$$+\frac{2\pi i}{h}\left(\Delta S + \frac{2}{A}(\nabla A \cdot \nabla S)\right) \qquad (11)$$

Substituting expression (11) into the Schrödinger Equation (7) and separating the real and imaginary parts, we now obtain two equations

$$\frac{(\nabla S)^2}{2m} + U - \frac{h^2}{8\pi^2 m}\frac{\Delta A}{A} \equiv \frac{m V^2}{2} + U + U_Q = \varepsilon \qquad (12)$$

$$\Delta S + \frac{2}{A}(\nabla A \cdot \nabla S) \equiv \nabla \cdot (A^2 \nabla S) = 0 \qquad (13)$$

Eq. (12) can be considered as a modified Hamilton-Jacobi equation for a particle moving with speed

$$\overline{V} = (1/m)\nabla S \qquad (14)$$

under the action of two potentials: the classical potential U and an additional potential U_Q.

$$U_Q = -\frac{h^2}{8\pi^2 m}\frac{\Delta A}{A} \quad . \qquad (15)$$

In addition the motion of the particle complies with condition (13)

Note that potential U_Q vanishes, that is $U_Q = 0$ when

$$\Delta A = 0 \qquad (16)$$

In this case the problem reduces to the classical problem of the motion of a material particles in external potential field U. Condition (16), however, means that in this case function $A(x,y,z)$ must be a harmonic function. Classical Liouville Theorem states that any bounded harmonic function on R^n is constant. Consequently, in order for additional potential U_Q to be equal to zero in unbounded configurational space, it is necessary that $A(x,y,z) \equiv const$. But in this case integral (4) diverges!

Thus, for any potential U the Schrödinger Equation (7), along with condition (5) and the substitution (9), has no solution that would coincide with any solution of a classical mechanics problem of the motion of a particle in the presence of external potential field U only, if motion occurs in unbounded space.

3.1. *The quantization of the orbital angular momentum*

If we require that the complex function $\psi(\overline{r})$ is single – valued, then by using the substitution (9) we extract from all the solutions of the Hamilton-Jacobi equation only those for which the orbital angular momentum is quantized. Indeed, consider a particle that is moving along any closed path $r = r(\varphi)$, $0 < \varphi \le 2\pi$. The condition for $\psi(\overline{r})$ to be a single-valued can be expressed in the form $\frac{1}{h}\oint dS = k$, $k = 0, \pm1, \pm2, ...$ where the integral is taken along this trajectory. Taken into account that $dS = \frac{\partial S}{\partial r}dr + \frac{\partial S}{\partial \varphi}d\varphi$, this condition can be represented in an equivalent form

$$\oint \frac{\partial S}{\partial \varphi}d\varphi = \oint \frac{1}{r}\frac{\partial S}{\partial \varphi}r d\varphi = m\oint V_\varphi r d\varphi = k h \qquad (17)$$

In a special case of the circular orbits with radius r, Eq. (17) takes the form: $2\pi m V r = k h$.

3.2. *Hydrodynamic Approach by Erwin Madelung*

Substitution (9) was introduced by Erwin Madelung . In 1926 after Schrödinger published his equation, Madelung published an article in which he discussed the hydrodynamic analogy between particles' trajectories and fluid motion. He substituted Eq. (9) into time-dependent Schrödinger equation and, utilizing the method described above, was able to obtain equations which in the case when functions A and S do not depend on time t explicitly are equivalent to the system of equations (12)-(13). If A^2 in Madelung's equations is interpreted as density $\rho = A^2$ of the vortex-free compressible fluid then Eq. (12) is equation of motion of this fluid and Eq. (13) is a continuity equation. The potential

$$U_Q \equiv p_Q = -\frac{h^2}{8\pi^2 m^2}\frac{1}{\sqrt{\rho}}\Delta(\sqrt{\rho}),$$ in this case, is

analogous to a pressure of unknown nature.

3.3. *The Quantum Potential Approach by David Bohm*

D. Bohm was a proponent of the causal interpretation of quantum theory and, therefore, existence of hidden variables. In 1952 he published two articles [2] in which an alternative to the Copenhagen interpretation was proposed, based on the notion of a particle acted on by a new kind of field. Bohm utilized the same substitution (9) as was previously used by Madelung and obtained system (12) - (13). However, Bohm's interpretation of the preceding system of equations had fundamental errors which are addressed bellow.

a) Bohm called U_Q "a quantum potential", because he assumed that in the classical limit ($h \to 0$) $U_Q \to 0$ and Eq. (12) takes a form of the Hamilton-Jacobi equation with a given potential U. This statement, however, is incorrect and is, unfortunately, found in the works of D. Bohm's followers as well as in many textbooks and even in Wikipedia.

Actually, U_Q has a finite value in the limit $h \to 0$. It can be proven by direct calculation of U_Q from the solutions of the Schrödinger equation for a hydrogen atom. As it is know, these solutions can be written in the form

$$\psi\,(r,\vartheta,\varphi) = A_{nlk}(r,\vartheta)\exp(ik\varphi)\ ,$$

$$n = 1,2,3,...,\infty\ ;\ l \le n-1\ ;\ k = 0,\pm 1,\pm 2,...,\pm l\ . \quad (18)$$

Hence, for $n = 1$ there is one solution

$$\psi_{100} = A_{100} = \frac{1}{\sqrt{\pi}}e^{-\frac{r}{r_0}}\ ,$$

$$U_q = -\frac{h^2}{8\pi^2 m_e}\frac{\Delta A_{100}}{A_{100}} = -\frac{h^2}{8\pi^2 m_e A_{100}}\left(\frac{d^2 A_{100}}{dr^2}+\frac{2}{r}\frac{dA_{100}}{dr}\right) =$$

$$= \frac{h^2}{8\pi^2 m_e r_0}\left(\frac{2}{r}-\frac{1}{r_0}\right) \quad (19)$$

where the Bohr radius r_0 is given by

$$r_0 = \frac{h^2}{4\pi^2 m_e e^2} \quad (20)$$

Substituting expression for r_0 into Eq. (19) we see that U_Q has a finite value as $h \to 0$.

A similar result can be obtained for any solution of the Schrödinger Equation. For instance, when $n = 2, l = 1, k = \pm 1$,

$$\psi_{21\pm 1} = \frac{r}{8r_0\sqrt{\pi}}e^{-\frac{r}{2r_0}}\sin\vartheta\,e^{\pm i\varphi}\ ,$$

$$U_Q = h^2/8\pi^2 m_e\left(1/r^2\sin^2\vartheta-2/r\,r_0+1/(2r_0)^2\right). \quad (21)$$

b) Bohm understood A^2 as the probability distribution of particles in a statistical ensemble of similar systems, and Eq. (13) as a continuity equation for the probability distribution. With this interpretation of A^2, in the deterministic equation of the particle's motion we have potential U_Q that itself depends on the particle's statistical characteristics. Besides that, experiments show that wave properties are characteristic of each individual elementary particle, not the ensemble.

c) Bohm inferred U_Q of having properties of nonlocality. He wrote: "*The quantum potential U_Q is different in many ways from classical potentials. The first key difference is that multiplication of the wave function by a constant does not change the quantum potential. The quantum potential can therefore still be large even when the wave function is small. This means that its effects **do not necessarily fall off with the distance**.... The quantum information potential has the new feature of **nonlocality,** implying an instantaneous connection between distant particles*".

But these Bohm's statements about quantum potential are unclear and even erroneous. Indeed if we choose some particular solution of the Schrödinger equation characterized by energy ε^*, define U_Q^* from that solution and find V^* according to Eqs. (9) and (14), the Eq. (12) becomes identity. For example, in the case of a hydrogen atom Eq. (12) leads to the identity

$$U_Q^* \equiv \varepsilon_n - \frac{m_e(V^*)^2}{2} + \frac{e^2}{r} \qquad (22)$$

where ε_n is the Bohr's energy level (6).

For example, when $n = 1$ we have

$\varepsilon_1 = -\dfrac{2\pi^2 e^4 m_e}{h^2}$, $V^* = 0$ (because the phase of the wave function is equal to zero) and identity (22) takes the form

$$U_Q^* \equiv \frac{e^2}{r} - \frac{2\pi^2 e^4 m_e}{h^2} \qquad (23)$$

It is evident from identity (22) that the additional potential U_Q (the quantum potential) is a nontrivial standard function on each of the trajectories obtained directly from the Schrodinger Equation. Moreover, Eq. (15) that defines this potential is valid on these trajectories only. However, outside those trajectories the additional potential U_Q is an unknown function.

3.4. Katel'nikov's approach

It is evident from the above that substitution (9) links Schrödinger Equation with classical mechanics. Naturally, questions arises: is there a more general substitution that provides a link between Schrödinger equation and classical mechanics? and how artificial is this link? The answers to these questions can be found in a work by the well known Russian physicist V.A. Kotel'nikov in [4].

Katel'nikov worked with the model of a quantum particle having both corpuscular and wave properties. He formulated the problem as follows: how can an elementary particle move according to the laws of classical mechanics if its probabilistic behavior is determined by the Schrödinger equation? In his study Kotel'nikov as a starting point used probabilistic interpretation of the ψ function. He proved that it is possible to obtain information about trajectories of a particle and velocities of the particle on these trajectories from the Schrödinger equation. As the result of his mathematical derivations, he concluded that the particle should move under the action of two forces: a classical force, defined by the potential U and a "quantum" force. If the general solution of the Schrödinger equation is represented in the following form $\psi = A\exp(i\beta)$ where $A(\overline{r},t)$ and $\beta(\overline{r},t)$ are real, then Katel'nikov's approach leads to formula (15) for potential energy of the "quantum" force. The equation for the particle's velocity will have the form: $V(\overline{r},t) = (h/2\pi m)\nabla\beta(\overline{r},t)$.

Unfortunately, the work of Katel'nikov has had little impact since a small number of copies of his work, not completed because of his death, were published in Russia and three chapters were published separately in the English version [4] of the Russion journal *Physics-Uspekhi (Advances in Physical Sciences)*.

Note, that under Katel'nikov's approach we assume that a particle can have a priori properties corresponding to noncommuting operators (for example trajectory and velocity of the particle on that trajectory). Therefore, this approach suggests existence of hidden variables.

4. The Schrödinger equation as a condition of stability

Consider the variational principle (3) that was used by Schrödinger in his derivation. As early as 1929 N.G. Chetaev, a well-known expert in the theory of stability, worked in University of Göttingen where he must have become familiar with the Schrödinger's work. In 1931, when Chetaev returned to the USSR, he published article [5] in which an attempt was made to justify the classical approach as well as to explain the variational principle (3) from the viewpoint of the theory of stability. He assumed that Schrödinger Eq. (9) under condition (5) extracted from all solutions of the Hamilton-Jacobi equation only those that satisfy the condition of stability. However, Chetaev was unable to obtain the Schrödinger equation exactly due to a flawed assumption that U_Q is the potential energy of small disturbing forces. Below we derive the Schrödinger equation using the Chetaev's approach, but correcting his mistake. Unlike Chetaev we start by introducing an unknown potential Φ, which is an operator dependent on a trajectory [6].

The Hamilton-Jacobi equation with a given potential energy U and additional potential energy Φ has a form

$$\frac{(\nabla S)^2}{2m} + U + \Phi = \varepsilon \tag{24}$$

where m is the mass of the particle. Eq. (24) was derived for a material particle; however, it is also valid for the motion of a center of mass of an extended object and also for the motion of center of mass with rotation about center of mass, if the energy of rotational motion is included in Φ.

Consider now the motion of a particle that it would have if small disturbing forces with potential energy W are present. Eq. (24) in this case takes the form:

$$\frac{(\nabla S)^2}{2m} + U + \Phi + W = \varepsilon \tag{25}$$

Of all the possible motions of a material system, we will consider only those for which the arbitrary constants of the complete integral of the Hamilton-Jacobi equation (integral Jacobi) have certain given values, and will call the collection of these motions a *packet*. Follow Chetaev's method, we assume that the influence of the perturbing forces on a packet at an arbitrary point is proportional to the density of trajectories A^2 at that point. For the packet consisting of stable trajectories this influence must be minimal. That is, the action of disturbing forces is relatively less for the packets for which

$$\int W \psi \psi^* d\tau \Rightarrow \min, \tag{26}$$

here ψ is determined from equation (9). While concurrently, taking the density A^2 to have the following condition

$$\int \psi \psi^* d\tau = \int A^2 d\tau = 1 \tag{27}$$

The integrals are taken over the entire volume of configuration space. Substituting the expression for W from Eq. (25) into Eq. (26), we obtain the following variational problem

$$\delta \int F \, d\tau \equiv \delta \int \left(\frac{(\nabla S)^2}{2m} + U + \Phi - \varepsilon \right) A^2 d\tau = 0 \tag{28}$$

Functions $S(\vec{r})$ and $A(r)$ realize extreme of the definite integral (28), if they satisfy the system of two Euler–Lagrange equations:

$$\frac{\partial}{\partial x}\left(\frac{\partial F}{\partial S_x'} \right) + \frac{\partial}{\partial y}\left(\frac{\partial F}{\partial S_y'} \right) + \frac{\partial}{\partial z}\left(\frac{\partial F}{\partial S_z'} \right) - \frac{\partial F}{\partial S} = 0 \tag{29a}$$

$$\frac{\partial}{\partial x}\left(\frac{\partial F}{\partial A_x'} \right) + \frac{\partial}{\partial y}\left(\frac{\partial F}{\partial A_y'} \right) + \frac{\partial}{\partial z}\left(\frac{\partial F}{\partial A_z'} \right) - \frac{\partial F}{\partial A} = 0 \tag{29b}$$

Assuming that operator Φ do not change under variation of S, we obtain Eq. (29a) in the form

$$2(\nabla A)(\nabla S) + A \Delta S = 0 \tag{30}$$

Eq. (30) coincides exactly with Eq. (13). Assume that operator Φ depends on A and does not depends on derivatives A_x', A_y', A_z', then condition (29b) can be represented in

$$\frac{(\nabla S)^2}{2m} + U + \Phi_o + \frac{\partial \Phi}{\partial A} - \varepsilon = 0 . \tag{31}$$

If on the trajectories that are solutions of the variational problem (28) (that is satisfy the necessary condition of stability) we take

$$\frac{\partial \Phi}{\partial A} = 0 \quad \text{and} \quad \Phi_o \equiv U_Q = -\frac{\hbar^2}{2m} \frac{\Delta A}{A} \tag{32}$$

than Eq. (31) coincides exactly with Eq. (12). It is obvious that the system of equations (30), (31) and (32) along with substitution (9) are equivalent to the time-independent Schrödinger equation. In order to prove this statement one should perform derivations with the separation of the real and imaginary part of the Schrödinger equation [refer to chapter 3], with the manipulation steps done in opposite order. Taking conditions (32) into account, allows the substantial nonlinear problem of calculating trajectory by means of operator Φ that itself depends on that trajectory to simply reduce to a linear Schrödinger equation.

This way, eigenvalues ε_n that are obtained from the time-independent Schrödinger equation are the ones that are extracted from solutions of Eq. (24), only those solutions that satisfy the following conditions:

(I) the orbital angular momentum of a particle moving along a closed trajectory **is quantized;**

(II) the motion of a particle that occur in agreement with these solutions **satisfy the necessary condition of stability (28);**

(III) additionally the potential Φ along those trajectories that satisfies the condition of stability (28), is equal to the so-called quantum potential U_Q.

Among these solutions (even within the same packet) there can be "extra" solutions that despite satisfying the necessary condition are unstable, and, therefore, are not realized in nature. Theoretically speaking, in order to find these "extra" trajectories we need to clarify the form of the potential Φ. This mysterious potential Φ, which is responsible for stabilization of the election's motion along the orbits, is a mathematical representation of some physical properties of «hidden variables».

What can be said about the unknown potential Φ on the basis of the proposed above derivation of the time-independent Schrödinger equation from the laws of classical mechanics?

a) Potential Φ and all its derivatives are continuous functions in the vicinity of the trajectories satisfying the necessary conditions of stability. On these trajectories and only on them Φ is equal to quantum potential U_Q. Here we see the main Bohm's and all his successors 's mistake; they assumed that potential U_Q is acting over entire configuration space although it is defined by Eq. (23) only on the trajectories that follow from solutions of the Schrödinger equation. On each such trajectory Φ is equal to a particular function U_Q, and off the trajectories Φ remains an unknown

b) The potential Φ depends on the form of the trajectory. Additionally, it is important to note that the velocity of each particle along the trajectory of a packet is a non-arbitrary value. Therefore, the potential Φ implicitly can depend on the velocity of the particle. It can be visualized on the following simple example: a body is moving uniformly along the bottom of a pool filled with ideal incompressible fluid. In this case in the body's coordinate system the pressure field in fluid depends on the speed of the body and the distance from the body to the bottom of the pool. If, for some reason, the speed of the body is not arbitrary but is a function of the distance, that would be an analogy to a packet of trajectories. Note, that nonlocality is present in this problem in the following way: the pressure in an ideal incompressible fluid is transmitted simultaneously over entire volume, that is, there is «*an instantaneous connection between distant particles*». However, no one talks about information potential in hydrodynamics. Nonlocality here, is merely a part of the model. In real fluids there exists a pressure wave – a "precursor" which propagates with speeds much greater than that of the moving body.

c) In general, there are two approaches to study the stability: one is to study the stability of a non-

perturbed motion with the addition of small perturbation forces (as seen above), while the other is to study the stability of a motion under variation of the initial coordinates and momenta with the same primary forces. The later, which is the most widely used one, is based on the Poincaré variational equations. It can be demonstrated using this approach that the Schrödinger equation is a necessary condition for stability of the motion under a variation of the initial coordinates only (conditional stability). Under this approach, it can be proven that for most situations the condition (30) splits into two conditions:

$$\Delta S = 0 \quad \text{and} \quad (\nabla A)(\nabla S) = 0. \quad (33)$$

It follows from the second equality that function A is constant on the trajectories that satisfy the necessary condition of stability. Taking this into consideration the first of the conditions (32) can be interpreted as follows: the decomposition into series of the potential Φ in the direction perpendicular to trajectory has the form

$$\Phi = \Phi_0 + \frac{\partial^2 \Phi}{\partial A^2}(A - A_0)^2 + o\left\{(A - A_0)^2\right\} \quad (34)$$

Following from (34) the force acting on the particle from an unknown field when the particle undergoes a small deflection from the given trajectory is analogues to Hooke's force. This force is the one that can stabilize the motion of a particle along the orbits corresponding to eigenvalues of energy. How can Hooke's law arise in this situation? It appears if we assume that structures (quasi-particles) are formed in the physical vacuum associated with the motion of particles. Stabilization of a particle motion can be visualized as follows: the particle remains on the surface of a "bubble" and any inward or outward deformation of the surface of the bubble causes an elastic restoring force.

In conclusion I would like to go back to the Madelung's hydrodynamic approach mentioned in the previous chapter. From the point of view of the theory developed here Madelung's approach is shown to have an analogy between an ensemble of trajectories which satisfy the necessary condition for stability and fluid currents.

5. Electron in the field of an atomic nucleus. Explanation of Rydberg formula

Now let us analyze the Schrödinger equation for the hydrogen atom {Eq. (4), $\kappa = h$} in the classical approach. Note here, that solutions of the Eq. (4) can be derived strictly mathematically without introducing any operators from quantum mechanics. The solutions can be written in the form $\psi\,(r,\vartheta,\varphi)=A_{nlk}(r,\vartheta)\exp(ik\varphi)$ (it can be found in textbooks). From these solutions one can obtain the velocity and the trajectory of the electron's motion, as well as the quantum potential U_Q. Indeed, the phase of the wave function (9) can be written as $2\pi S/h=k\varphi$, where $k=0,\pm1,\pm2,...,\pm l$; $l\le n-1$. As known, ∇S in spherical coordinates has the form: $\nabla S=\dfrac{\partial S}{\partial r}\bar{i}_r+\dfrac{1}{r}\dfrac{\partial S}{\partial\vartheta}\bar{i}_\vartheta+\dfrac{1}{r\sin\vartheta}\dfrac{\partial S}{\partial\varphi}\bar{i}_\varphi$. The component of the velocity along the radius $\bar{i}_r$ is zero, because $\dfrac{\partial S}{\partial r}=0$; the component of the velocity along $\bar{i}_\vartheta$ is also equal to zero because $\dfrac{\partial S}{\partial\vartheta}=0$. The only nonzero component of the velocity is the component along $\bar{i}_\varphi$. By substituting the above expression into the formula (12) for the velocity of the electron on the trajectories obtained from the Schrödinger equation we obtain

$$\bar{V}=\frac{1}{m_e}\nabla S=\frac{hk}{2\pi m_e r\sin\vartheta}\bar{i}_\varphi \qquad (35)$$

From the formula (35) it can be seen that only two situations are possible: either the center of mass of the electron is at rest ($k=0$) or is moving along a circular orbit lying in a plane parallel to the xy plane with the center on z axis (the above reasoning is applicable to any coordinate axis). Since the motion on these circular trajectories satisfies only the necessary condition of stability, among these trajectories there can be trajectories that are not stable.

5.1. *Bohr orbits*

Mathematical derivations presented below are valid for any circular orbits; however, to simplify our calculations without loss of generality we will consider the circular orbits with the center at the origin. It follows from Eq. (35) that the speed of an electron along the circular path with the centers at the origin can be represented as

$$\bar{V}=\frac{\nabla S}{m_e}=\frac{hk}{2\pi m_e r}\bar{i}_\varphi \qquad (36)$$

If all of the above orbits with various radii exist in nature (that is the orbits would not only satisfy the

necessary but also the sufficient conditions for stability), then from the Newton's second law it would follow directly that

$$\frac{m_e V^2}{r}=\frac{h^2\,k^2}{4\pi^2 m_e r^3}=F, \qquad (37)$$

That is, there exists an unknown force F acting on the particle for any given radius; in this case the external potential U would be irrelevant. There are, however, certain orbits r_B at which force F coincides with the Coulomb force. It can be easily verified that these orbits are Bohr orbits

$$r_{B_k}=\hbar^2\,k^2\,/\,m\,e^2 \qquad (38)$$

Thus Bohr orbits are the solutions of the Schrödinger equation for the hydrogen atom under the classical approach. Moreover, from all of the possible solutions of the Schrödinger equation, the Bohr orbits are the ideal candidates for being stable.

5.2. *The precession of the electron's spin in an atom*

For Bohr orbits the following motion integral is valid

$$(1/2)\,m_e V_k^2-e^2\,/\,r_{B_k}=\varepsilon_k \qquad (39)$$

However, the Schrödinger equation gives the following generalized motion integral for Bohr's orbits

$$\frac{m_e V_k^2}{2}-\frac{e^2}{r_{B_k}}+U_Q=\varepsilon_n \qquad (40)$$

where ε_n are the energies on the Bohr orbits, determined from Eq. (6).

Comparing the above two expressions, for Bohr's orbits we obtain

$$U_Q=\varepsilon_n-\varepsilon_k .$$

Thus, in Eq. (40) along with the term that characterizes the energy of the center of mass, there is another term, which determines $U_Q=\varepsilon_n-\varepsilon_k$. This brings up a question: under which physical assumptions are the motion integral (40) possible? The motion integral in form (40) is possible if the motion of an object can be represented as a superposition of two motions: the motion of its center of mass and the motion about the center of mass. The extra term $U_Q=\varepsilon_n-\varepsilon_k$ in this case is the energy associated with the motion about the center of mass (for example, the precession of the electron's spin). We will discuss this in detail below.

The equations of motion of the object's center of mass and the equations of motion about the center of mass may not be independent, and because of that they cannot be analyzed separately. In such case the Hamilton's principal function, as well as generalized motion integral, depends on variables corresponding to both rotational and translational motion. However, there might be partial solutions in which generalized motion integral splits into two independent parts: the motion of center of mass and the motion about the center of mass. The motion integral (40) is an example of such case.

Suppose that the motion about the center of mass is the precessional motion of the electron's spin. We will describe spin precession as precession of a classical gyroscope, because there is a preferable direction in this problem (which is determined by the plane of an orbit and the normal to this plain). It follows from gyroscopic theory that for precessional motions of fast gyroscopes' which are limited to small precessional angles, the generalized motion integral contains energy E corresponding to the precessional motion only:

$$E = U_Q = \varepsilon_n - \varepsilon_k \qquad (41)$$

The energy E numerically equals the product of the intrinsic angular momentum and the angular frequency of precession ω_e

$$E = h\omega_e / 8\pi \qquad (42)$$

Note that in the case of precessional motion, energy depends on the frequency of precession and not on the square of the frequency, as in the case of wave motions.

Thus on the Bohr orbit potential U_Q ("a quantum potential") equals to the energy of the precessional motion of the electron's spin. It can be seen from Eq. (41) that the formula for the energy of the precessional motion is, in fact, the Rydberg's formula.

5.3. The Superfluid Physical Vacuum

Up to now we did not introduce any hypothesis besides the hypothesis of existence of "hidden variables" like a physical field. If we make an assumption that this field as a material substance is in various aspects like superfluid He-3 [7], then we can provide an explanation of the natural frequencies of the atom. Indeed, in superfluid He-3 structures (quasi-particles) like homogeneous precessing domains are observed, where all spins of the fluid particles precess with the same frequency and phase. These quasi-particles have spin proportional to h. In the presumed superfluid medium (the superfluid aether) the electron's motion can create structures similar to such domains. Then from the law of conservation of the angular momentum the torque acting on the electron's spin that causes its precession

with frequency ω_e must be numerically equal to the torque acting on the vacuum's spin structure; that is the following equality holds

$$|\varepsilon_n| = \omega_e \cdot h / 4\pi = \omega_p \cdot h / 2\pi \qquad (43)$$

where ω_p is the angular frequency of the precession of the vacuum structure's spin. Note that Eq. (43), in essence, is a new interpretation of the De Broglie's "law of phase harmony": a term ε / h in the expression of the non-stationary component of the wave function phase represents on the one hand the precession frequency of the electron's spin, and on the other hand is a frequency of coherent precession of the superfluid vacuum particles [8].

It follows from Eqs. (41), (42) и (43) that the angular frequency of the precession of the vacuum structure's spin ω_p is calculated from Rydberg formula $\omega_p \cdot h / 2\pi = \varepsilon_n - \varepsilon_k$. This approach is consistent with the gyroscopic models of emission – absorption of a resonance photon.

Conclusion. The classical approach presented above gives us a strong mathematical base with which to suggest that the electron's spin in an atom is precessing. It also opens a new direction to the study of the nature of "hidden variables".

Acknowledgments.

I would like to thank Nadia Lvov, Vladimir Bychkov and Dmityi Kalantarov and for their help in preparation of the article and for valuable comments.

References

1. Don Howard, *Studies in History and Philosophy of Science, Part A*, **16 (3),**171 (1985)
2. D.Bohm , *Phys. Rev.*, **85 (2),** 166 (1952).
3. E. Schrödinger, *Annalen der Physik*, **79**, 361(1926).
4. V.A.Kotel'nikov, *Phys.- Usp.*, **52**, 185 (2009).
5. N. G. Chetaev, Stability of Motion. Proceedings in Analitical Mechanics (USSR Academy of Science Publishing, Moscow, 1962) [in Russian].
6. N. Sotina, *Physics Essays*, **27**, 321 (2014).
7. L. Boldyreva & N. Sotina, *Physics Essays*, **5**, 510 (1992)
8. N. Sotina, Reports of the 16th RCCNT&BL, Dagomys, June 2009, Moscow, 172 (2010)

Possible Electrical Shaping Forces for Saturn's Circumpolar Hexagon

BARRY SPRINGER

11450 N. Meridian Rd
Falcon, Colorado 80831

E-mail: bspringer@skybeam.com

Images of the north polar region of Saturn reveal dynamic but persistent cloud patterns in the shape of a nested set of hexagons. Investigators have noted the presence of numerous towering storms outside the region of the hexagon channel but have offered no scale appropriate theory on how these storms could self-organize and maintain a stable hexagonal structure in which storms are generally absent. Saturn is known to be electrically active as evidenced by circumpolar auroral activity. Auroras are produced by inflow of energetic particles that constitute electrical currents. These currents could produce forces to shape flow of naturally occurring atmospheric motion. A preliminary investigation and conjecture is offered that models dual spiraling ionic vortex flows that could influence movement of surface winds, suppressing storms in a mid-high latitude band to create the observed nested hexagonal shapes.

1. Introduction

Fig.1 shows an image of Saturn's hexagon, taken by the Cassini spacecraft in July 2013 from about 973,000 km. from the planet. The image has a scale of 58 km per pixel and faces the sunlit side of the rings, from about 33 degrees above the ring plane. The hexagon spans about 30,000 km across. [1]

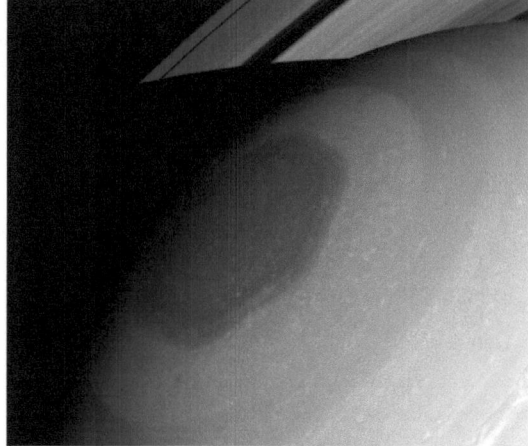

Fig. 1. Cassini image of Saturn's polar hexagon.

Planetary probes have for several decades observed the concentric hexagonal patterns in storms at the polar region of Saturn. But, investigators have had great difficulty proposing a mechanism by which such storm patterns could self-organize and maintain an organized pattern over decades as has been observed.

Cloud tracking wind analysis by Godfrey found that wind velocities in the region of the hexagon, centered between 74 and 78 degrees latitude reached peaks over 100 m/s. [2] Allison has proposed that the hexagon feature is a stationary Rossby wave created by shear

forces and trapped by the strong relative vorticity gradient of the wind stream itself. [3]

Fig. 2 graphs observations of zonal winds on Saturn revealed in the 5.1 um thermal imagery. These wind patterns are described by Baines et al as two distinctive cloud "tracks" separated by a nearly cloud free track, extending over approximately 3^0 of latitude centered about 76^0 latitude. [4]

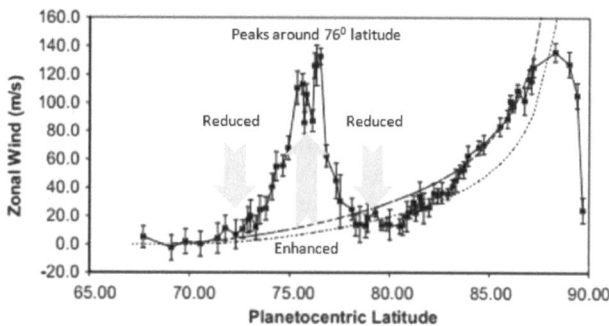

Fig. 2. High latitude wind profile on Saturn.

Fig. 3 shows the closest that laboratory efforts have come to creating a physical process to explain the hexagonal structure. This experiment involved shear forces created by differential fluid flow, the results of which correlate only very roughly with the sharply defined hexagonal feature on Saturn. [5]

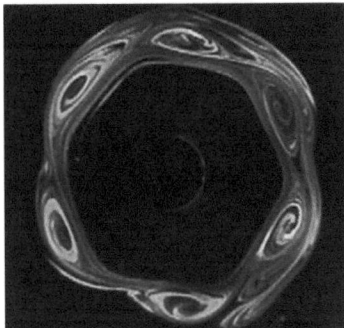

Fig. 3. Laboratory experiment.

None of the researchers have yet considered or evaluated any mechanism external to Saturn. It may be possible that external electrical forces are shaping the movement of atmospheric winds and affecting creation of storms. We will now explore that possibility.

2. Saturn's Aurora.

Auroral activity has been seen surrounding the poles of Saturn, caused by interaction of highly energetic particles interacting with gases in the upper atmosphere. Fig.4 shows Saturn's aurora in ultraviolet wavelengths. [6]

Fig. 4. Saturn's ultraviolet aurora.

3. Electrical Vortices.

Energetic particle flows can also take the form of very powerful electrical currents in space. Where current densities are high, parallel current streams are also known to twist around one another and form rotating vortices of electrical current known as Birkeland currents. Fig. 5 depicts a pair of Earth impinging counter rotating Birkeland currents that have been detected by NASA THEMIS spacecraft above the polar regions of Earth. [7]

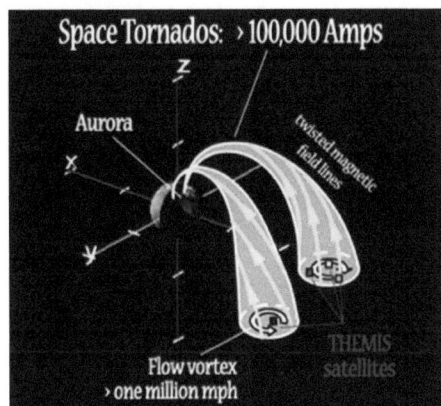

Fig. 5. Birkeland currents detected by NASA THEMIS satellites.

It is reasonable to assume that such electrical vortices may also be impinging upon the polar regions of Saturn.

Where such electrical vortices intersect a planet, the rotation of the planet would cause the intersection to form a circular pattern at the latitudes of the intersections. [8] But, the intersection pattern will be more complex than a simple circle, as the individual current filaments within each of the vortices are offset from the central axis and may also be rotating. The intersection of the vortices on the Earth would termed a roulette.

4. Mathematical Roulettes.

The pattern of a point offset from center of a circle rotating within or about another circle is mathematically described as a roulette. The pattern is more specifically described as a hypotrochoid or epitrochoid depending on positioning of the moving circle with respect to the base circle. [9] [10]

The trace patterns are influenced by the offset of the tracing point with respect to the center of the moveable circle as shown in Fig. 6 and Fig. 7. [9] [10]

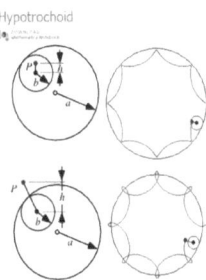

Fig. 6. Epitrochoids.

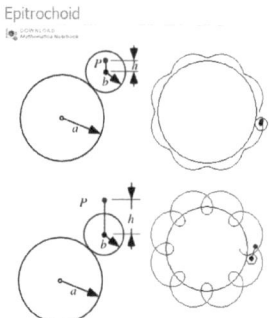

Epitrochoid

Fig. 7. Hypotrochoids

5. Mathematics of the Hypotrochoid.

The pattern of the hypotrochoid is described by only two parameters: "n" represents the wave number and "m" represents the degree of modulation of the waves about the circle.

As shown in Fig. 8, the equations that describe the hypotrochoid are a combination of scaling factors and sines and cosines, reflective of the originating mechanisms being a primary circle and a secondary constrained rotating circle. [9]

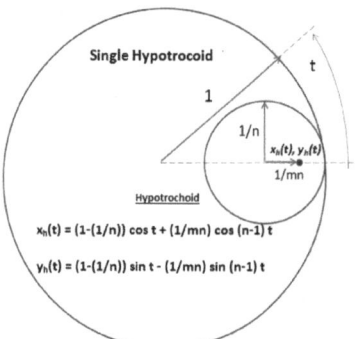

Fig. 8. Mathematics of the hypotrochoid.

To produce a hexagon, the wave number must be set to six and the modulation number may be varied as necessary to achieve a match to the observed nearly straight sides of the hexagonal shape.

Fig. 9 depicts a hypotrochoid created with a wave number of six and also a modulation number of six that produced the best expression of a nearly straight sided hexagon.

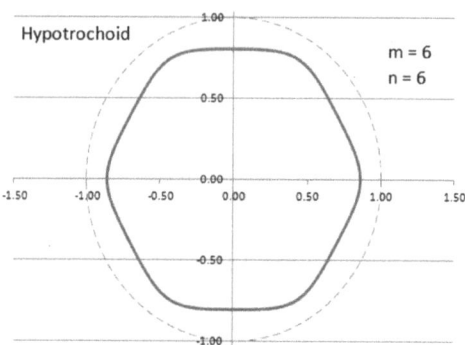

Fig. 9. Pattern produced by hypotrochoid of m=6, n=6.

6. Modeling Rotating Paired Vortices.

The mathematics of a hypotrochoid tangentially paired with an epitrochoid is shown in Fig. 10. The combined roulettes will trace out a concentric pair of figures. To achieve the same polygon shape, the wave numbers must be identical, but the modulation factors could be different. [9] [10]

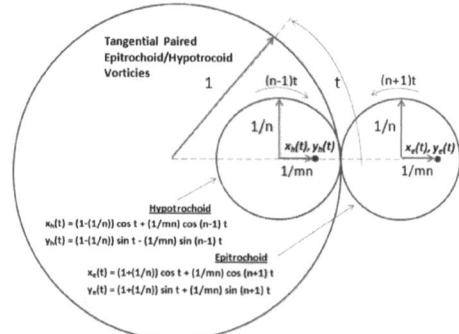

Fig. 10. Mathematics of tangential paired epitrochoid/hypotrochoid.

It is not necessary that the two figure generators rotate exactly tangentially, as the same figure could also be produced from opposing paired vortices as shown in Fig. 11.

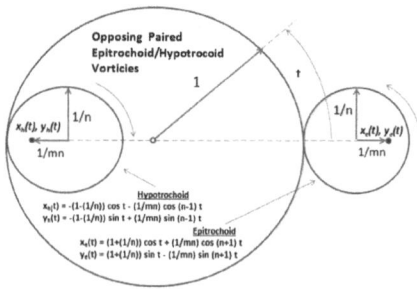

Fig. 11 Mathematics of opposing paired epitrochoid/hypotrochoid.

When a hypotrochoid is paired with an epitrochoid of identical parameters for wave number and modulation factor, a pair of nested hexagons is produced as shown in Fig. 12. It is interesting to note that the spacing between the nested hexagons is not an independent variable, but is determined uniquely by the modulation number chosen to replicate the smooth wall structure seen in Saturn's hexagon.

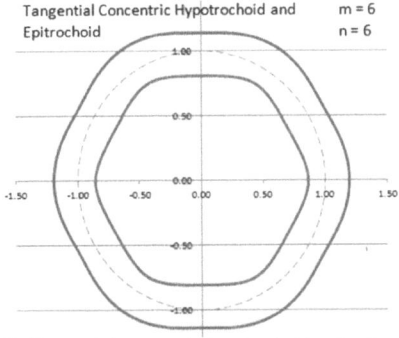

Fig. 12. Tangential concentric hypotrochoid and epitrochoid.

7. The Saturn Polar Hexagon.

Fig. 13 shows image an image of Saturn's polar hexagon collected by the NASA Cassini spacecraft. This is one of a series of images taken at the highest resolution available. [11]

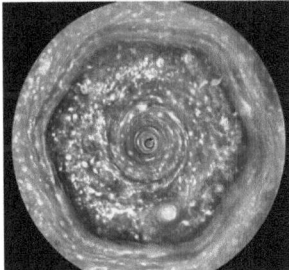

Fig. 13. Saturn's hexagon.

Fig. 14 shows an image made from data obtained by Cassini's visual and infrared mapping spectrometer in the 5-micron wavelength of radiation. Viewed in infrared, we see that the white banded wall structure area is brighter than the interior. White indicates clear areas that allow infrared emissions from the surface of Saturn to be seen.

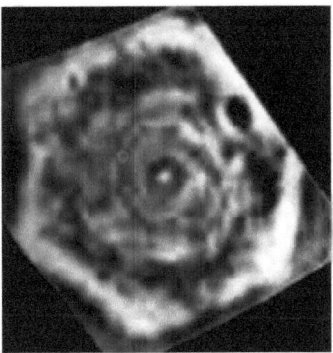

Fig. 14. Cassini infrared image (5 micron) of Saturn's hexagon.

We see that storm activity is generally confined to the interior of the inner and exterior of the dual hexagon wall, where the darker colors indicate reflected light from storm tops towering up to 25 km high. [12]

8. Shaping the Hexagon.

We began by suggesting that extra planetary forces could be involved in shaping the north polar hexagon on Saturn. We have posited that electrical current vortices swirling about the northern pole of Saturn could produce well defined polygons that can be mathematically described.

Fig. 15 shows the overlay on Saturn's polar hexagon of the posited electrical current shaping traces that may be created by paired current vortices intersecting a rotating planet. We find a remarkable correlation between the dual trochoid traces and the hexagonally shaped exclusion channel within which high linear winds exist and cyclonic storms are generally absent.

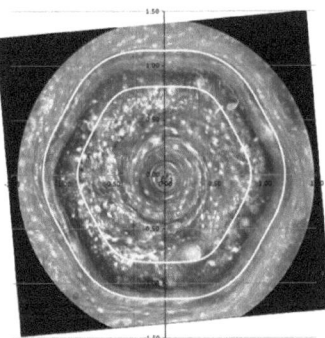

Fig.15. Possible shaping process overlaid on Saturn's hexagon.

9. Atmosphere of Saturn.

Saturn's atmosphere is composed of 96.3% hydrogen H2 and 3.25% Helium. Trace compounds include 0.45% methane (CH4), 0.0125% ammonia (NH3), 0.0110% Hydrogen Deuteride (HD) and 0.0007% ethane (C2H6). [13]

Clouds are composed of ammonia ice, water ice and ammonium hydrosulfide. The surface pressure is estimated to be greater than 1000 bars. Temperatures are -139 C at 1 bar and -189 C at 0.1 bar. Wind speeds up to 150 m/s exist above 300 latitude. [13]

10. Rotational Dynamics.

A day on Saturn is 10.656 hours long. The hexagon appears to rotate at the same rate as the planet, indicating it is stationary with respect to Saturn's core.

The hexagon displays a distinct channel bracketing +/- 2 deg of 76 deg. north latitude. If the shaping vortices intersect at fixed locations, the rotation rates would be the wave # divided by the rotation period (6/10.656) which would be 0.563 cycles/hr or 15.6 milliHz. [8] [13]

Fig. 16 details the rotational dynamics that would be postulated to exist if the vortices were impinging on Saturn at a fixed location Saturn centered about the 76^0 latitude level.

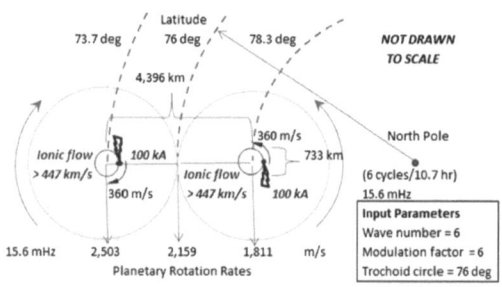

Fig. 16. Postulated rotational dynamics of dual vortices.

If the vortices were indeed impinging at a fixed location, then planetary rotation would carry the atmosphere through the impingement region, resulting in shaping effects in various regions where the atmospheric movement could be bucked, blocked and channeled as shown in Fig. 17.

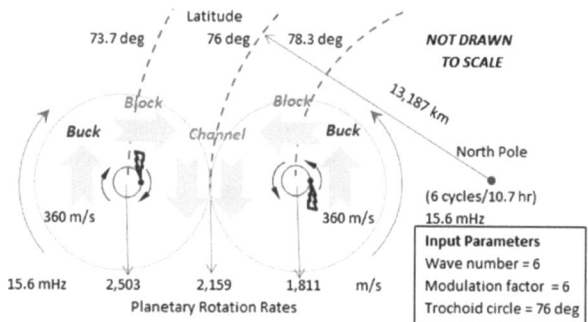

Fig. 17. Postulated effects shaping atmospheric motion.

The effects of such atmospheric flow shaping could be a factor influencing or channeling the latitudinal zonal flows measured in the 4^0 band centered about 76^0 latitude, as shown in Fig. 18. [4]

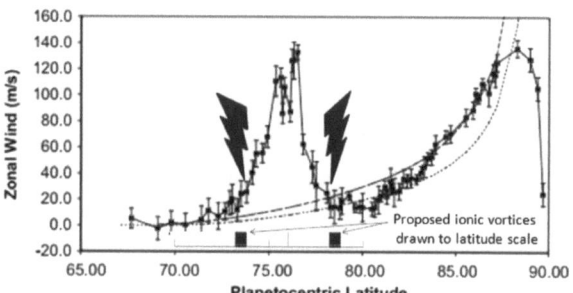

Fig. 18. Postulated influence on zonal winds patterns by vortices.

11. Conclusions.

Infrared imagery of Saturn reveals a zonal band in the 74-78 degree high latitude region in which high speed winds are found, which is also generally free of storms and the boundaries of which exhibit the shape of concentric hexagons.

A pair of highly energetic counter rotating ionic vortices has been discovered by NASA THEMIS satellites to be impinging on polar auroral regions of Earth and it is reasonable to presume that such vortices may be similarly impinging upon Saturn.

A pair of rotating vortices intersecting a rotating sphere may produce a dual walled hexagon traces described by concentric epicycloid and hypocycloid figures defined by a single wave number and a single modulation factor.

Shaping of Saturn's hexagon by external electrical forces offers a more refined fit to the observed phenomena than local aerodynamic wind shear effects associated with differential flows.

12. Conjecture.

Based upon the foregoing conclusions, we offer the following conjecture concerning the source and process by which the circumpolar hexagon on Saturn is created and sustained.

A pair of strong space Birkeland currents may be impinging upon Saturn's auroral region, directing atmospheric flow into a storm-free channel defined by rotational motion of impinging currents in epitrochoid-hypotrochoid cycles forming a hexagon.

13. Further Investigation.

This preliminary investigation constitutes initial steps in an inductive logic process. [14]

Thus far, we have presented observations of an unexplained natural phenomena and discovered correlations between the observed phenomena and a mathematical defining process. These steps have demonstrated observation and correlation and led to a conjecture, but we have not yet demonstrated causality.

Not yet evaluated is the presence around Saturn of a complex magnetic field with dipole strength of .21 gauss, dipole tilt within 1^0 of rotation axis and dipole offset of .04-.05 Rs northward. The central magnetic field is more complex than a dipole as quadrapole and octapole moments have also been observed.

It is unlikely that electrical vortices impinging Saturn just happen to exhibit an uninfluenced exact wave number of six and consistently impinge about 76 degrees latitude. It is more likely that there exists a resonant coupling mechanism between the Birkeland currents and Saturn's complex magnetic field that stabilizes the hexagon structuring forces at that latitude and disciplines the wave number.

Further research into these phenomena should be focused on discovering and modeling electromagnetic coupling mechanisms that would stabilize the latitude band and wave number of the conjectured impinging current vortices.

References.

1. NASA/JPL-Caltech/Space Science Institute, Cassini Spacecraft Image, July 2013, http://www.space.com/27392-saturn-hexagon-vortex-nasa-photo.html.
2. D.A. Godfrey, Icarus 76, 335, 1988.
3. M. Allison, A Wave Dynamical Interpretation of Saturn's Polar Hexagon, Science 247, 1061-1063, 1990, http://pubs.giss.nasa.gov/abs/al04100j.html.
4. Baines et al, Saturn's north polar cyclone and hexagon at dept revealed by Cassini/VIMS, Elsevier, Planetary and Space Science 57 (2009) 1671-1681, http://www.elsevier.com
5. Emily Lakdawalla, Saturn's hexagon recreated in the laboratory, http://www.planetary.org/blogs/emily-lakdawalla/2010/2471.html, 2010.
6. NASA, Hubble site gallery, http://hubblesite.org/gallery/album/solar_system/pr1998005a/ image of January 7, 1998.
7. Robert Sanders, THEMIS mission tracks electrical tornadoes in space, U.C. Berkley News, http://www.berkeley.edu/news/media/releases/2009/04/23_keiling.shtml, 2009.
8. Godfrey, D. A. (March 9, 1990), The Rotation Period of Saturn's Polar Hexagon. Science 247 (4947): 1206–1208.
9. Wolfram, Mathworld, Hypotrochoid, http://mathworld.wolfram.com/Hypotrochoid.html
10. Wolfram, Mathworld, Epitrochoid, http://mathworld.wolfram.com/Epitrochoid.html .
11. NASA/JPL-Caltech/SSI/Hampton University, NASA's Cassini Spacecraft Obtains Best Views of Saturn Hexagon, http://www.jpl.nasa.gov/news/news.php?release=2013-350.
12. NASA Cassini image of Saturn, NASA/JPL-Caltech/Space Science Institute, Looking Down on the Hexagon in Infrared, http://photojournal.jpl.nasa.gov/catalog/PIA17654
13. David Williams, NASA Goddard Space Flight Center, Saturn Fact Sheet, http://nssdc.gsfc.nasa.gov/planetary/factsheet/saturnfact.html.
14. David Harriman, The Logical Leap: Induction in Physics, New American Library division of the Penguin Group, New York, NY, pp 184-185, 2010.

GRAVITATIONAL ANOMALIES AND THE "ALLAIS EFFECT" FOUND IN ITALY DURING ECLIPES AND STRONG EARTHQUAKES

VALENTINO STRASER

Independent Researcher, Strada dei Laghi, 8 Terenzo-Parma, 43040, Italy

This paper presents data and gravitational anomalies measured during eclipses of the Sun and Moon and strong earthquakes. The measurements were made using a simple pendulum gravimeter, created by Dr. Mario Campion, which differs substantially from those normally used: the first type with a constant length spring, the second able to measure absolute gravity by use of a free-fall mechanism, and the third working with a sensitive electromagnetic suspension balance. During eclipses of the Sun and Moon, this instrument revealed strong anomalies in the signal that did not appear in the data measured by other gravimeters, but which do agree with the "Allais Effect". Moreover, the instrument can measure seismic shocks, even at great distances, providing graphs of positive semi waves that indicate variations in gravity at very low frequencies. In the partial solar eclipse of 4 January 2011, the gravimeter showed a rising peak in gravity, equal to 16 millionths of g, on the theoretical three star alignment and the passage of the Moon on the meridian. Some anomalous behaviour was also recorded during the eclipse of the Moon on 21 December 2010. The data show a variation (increase) in gravity of around 30 millionths, equal to a variation of 20 millionths of g, and corresponding to around 45 milligals. Instead, during the catastrophic earthquake in Japan (11 March 2011), the gravimeter recorded the event with low frequency signals and gravity diminished by 40 millionths of g, with only a half wave reduction.

Introduction

There should be a premise before submitting the data to the local gravity, in the town of Rovigo (Fig.1) coordinates (Latitude and Longitude 45.07 N. -11.778 E). The instrument used in this case for the monitoring of the gravitational field differs significantly from those normally used, which are essentially of three types: the first is a spring with constant length, the second is able to measure the absolute gravity, said also to fall, and the third one that works with the sensing element electromagnetic balance, and other the technologies (Imanishi et al., 2004). The instrument created by Dr. Campion deviates significantly from the three types described above, but which may have advantage aspects of measurement compared to them.

Figure 1. Index map showing the study area in a red point.

These basically refer a gravity meter that takes full advantage of the simple pendulum, when optimizing its operation with advanced technologies.

The average value of gravity measured in a defined time interval, the tool gets it, with extreme precision by timing the time it takes to make the oscillator 1000 or 100 or 10 oscillations, and then dividing the value for the same number of oscillations. When determining the extent of 1000 oscillations of **g**, the figure is averaged over nearly half an hour and at the end of the day, it creates a graph with 50 intervals on the abscissa and the same with their values on the ordinate.

The variations of the average period are inversely proportional to the variations of gravity g, in the range in which the instrument has been carried out the measurement, as is apparent from the formula of the simple pendulum. In the case in which one operates with 100 or 10 oscillations, the measures are obtained by intervals ten hundred times shorter and therefore with a capacity for analysis of phenomena very detailed. A point of great advantage of this instrument, according to Dr. Campion, is to sum all the values of the gravity that they have acted on the oscillator, one after another. Practically, the entire pattern of the force of gravity, punctuated by the oscillator, comes into play in the evaluation of its mean value, also performing a data compression, the number of which can be easily used becomes limited.

Turning to 100 measures the initial interval starting begins to make itself felt, in each case represents only thirty percent of the total interval. For 10 measures the measuring interval is reduced to ten percent, and in any case much wider of other measuring systems of **g**. To complete the description of the device, it is specified that is controlled by a computer which operates in an automatic way the operating cycle. The graphs are plotted daily occur with changes in the average period as an index of **g**, and are much larger than indicated values of gravity measured by the gravity-type spring or fall. In our case, the period may vary from 5 or 6 millionths, these variations are much larger than the values of g measured with other gravimeters.

This amplification could be a result of this measurement method, which regards a summation of repeated values and also for the fact that in this case we use a sensor which in addition possesses an angular momentum horizontal, that is perpendicular to both the plane of oscillation to the direction that **g**.

The graphs do not show daily and a regular basis curve of tidal forces, but over the cycle, there are two significant events in which they interpret the tidal forces in a clear, graphical plotting of similarity with those of the tidal forces, always with great amplification of the phenomenon (Straser, 2010).

Variations between gravitational and tidal force

The daily analysis of the graphs has allowed to deduce that during the lunar month there are two critical moments, one in the vicinity of the new moon (generally more pronounced), but also of the full moon, in which the trend of gravity recorded by the gravimeter agrees with the forecast data of the tides of the Adriatic Sea. The (Fig.2) shows the trend of the gravity February 2, 2011, the day preceding the new moon. The recording was made with the gravity meter set on the 100 measures recorded at intervals of 4 minutes. The trend line (black trace labeled), which represents the average of the measured data shows a clear trend that looks like the tides of the Adriatic Sea, with the maximum at the passages of the Sun and the Moon over the meridian.

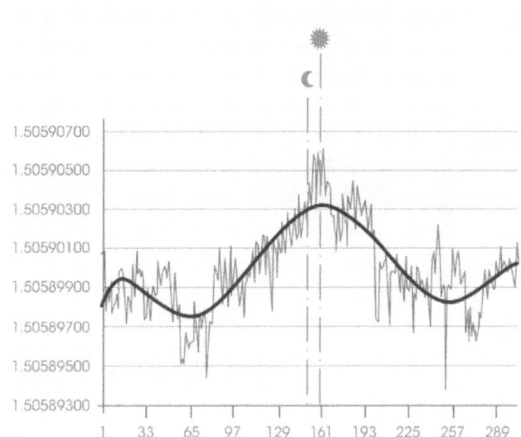

Figure 2. Trend of gravity February 2, 2011, the day preceding the new moon

If this trend is compared with the graphs of tidal forces, measured with other gravimeters, it appears evident the enormous amplification of tidal forces highlighted by this type of gravimeter.

In fact, the variation between maximum and minimum of the tide, which is deduced from the graph, is of 5.5 millionths of **g**.

Allais Effect

Allais Effect consist in an anomalous precession of the plane of oscillation of a pendulum. Maurice Allais saw for the first time this effect in 1954 and 1959 during the solar eclipses.

During the solar eclipses there is an alignment among the Sun, Moon and Earth and the rotation of the plane of oscillation of the pendulum change and its linked to the rotation of the Earth.

The rotation of the plane of oscillation of the pendulum from clockwise, counterclockwise becomes for the duration of the eclipse (Kokus, 2013).

- In the Northern Hemisphere, a pendulum should rotate clockwise.
- At the very beginning of an eclipse, the pendulum's rotation will deviate towards counterclockwise.
- Afterward, it returns to clockwise rotation.
- It has been observed during both solar and lunar eclipses.
- It has been measured on the side of the earth opposite the eclipse.
- Has been measured with a torsion pendulum, a stationary pendulum and a torsion balance pointer simultaneously at sites 300km apart.

- A.E.Pugach estimates that the angular acceleration of a body under the influence of the Allais Eclipse Effect would be 4.4x10-4 deg/s.

Instrument

The gravimeter has a device which is independent from barometric pressure variations and a pendulum with low expansion rods to limit errors due to thermal dilatation. The oscillator with a position finder able to produce a very precise synchronism signal has no electromagnetic interference and is connected to an electronic clock which is precise to the eighth-ninth significant figure. This system is controlled by a calculator. In one day, about 52 values of the Earth's gravitational field are obtained and data continue to be collected between one measurement and the next thanks to being recorded on a disk. The relative error over 1000 measurements is 0.000000089.

Discussion

Eclipses

1-The first case: partial eclipse of the Sun on 4 January 2011. As noted by Dr. Mario Campion, It appears evident that in the time interval between the theoretical alignment of three celestial bodies and the passage of the moon at the meridian, the gravimeter has detected a peak rising of gravity, equal to 16 millionths of **g** (Fig.3).

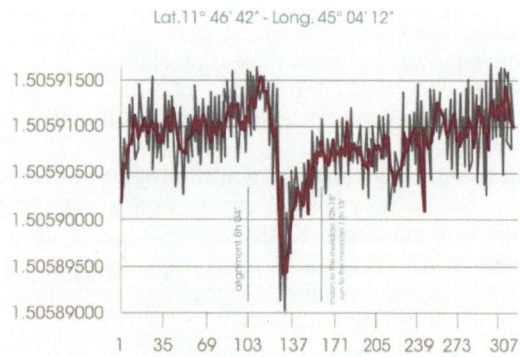

Figure 3. Eclipse of the Sun on January 4, 2011

It is a given instrumental obtained in conditions of normal operation, since the tool has operated in all tranquility, without external influences. As additional data, and as part of the assessment, it should be noted that there were no anomalies in the tides of the Adriatic Sea, during the time of the eclipse.

Some research institutions and universities have submitted their comments, which, during total solar eclipses were carried out to monitor the severity, but did not detect abnormalities details, but only modest decreases.

An objection was raised by the researchers was that the gravitational anomaly could depend on the aquifer underlying influences about the position of the gravimeter (Contadakis and Asteriadis, 2001; Tanaka et al., 2006))

Probably, however, it is a different sensitivity of the sensor which has detected the phenomenon shown in the graph, compared to the gravimeter to fall, which instead focuses on the vertical component of gravity. In this context, noted Dr. Mario Campion, it is worth remembering that even Maurice Allais noticed abnormal behavior of the Foucault pendulum during solar eclipses. In this regard, however, it is hypothesized that there may be a decrease in intensity of the Earth's magnetic field due to the eclipse of the sun.

2-The second case: total Eclipse of the Moon, December 21, 2010. This graph (Fig.4) shows in abscissa the number of measuring intervals, each of about 28 minutes and in ordinate the value of the period detected by the oscillator which presents with an originality even more marked of the preceding graph, relative the eclipse of the sun and therefore not in agreement with the findings obtained with other instruments.

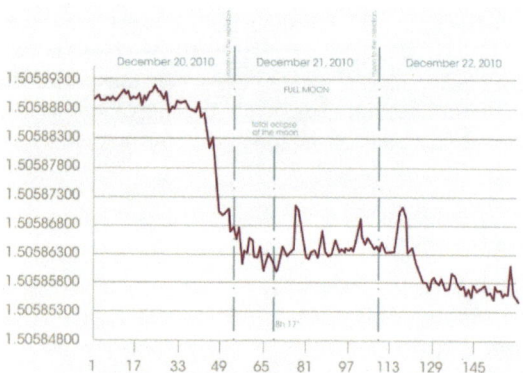

Figure 4. Total eclipse of the Moon on December 21, 2010

The data deducted from the graph reveal an increase in the level of gravity of about 30 millionths of the period, equal to a change of 20 millionths of g corresponding to about 45 milligal.

At this point, we should question the instrumental data gravimeter, which is also in this case it would be difficult to accept, because the trend of the graph looks anything but equivocal. The instrumental data show it as a tensile stress to an increase in the level of g, which starts six hours before the passage of the moon to the meridian and is completed by the occultation of the moon during the eclipse. That this is random, it is very strange. Observing the graph is better note a similarity with the other graph relating to the eclipse of the Sun, and that is that in this case also the variation of gravity takes place with an increase and not a decrease, even if not then return as another If its primitive values.

What can we say about this?
We can say that the tidal forces can be associated to various causes concatenated, as modest deformations of the soil, micro-displacements of the center of gravity of the Earth and other yet, which at the time to escape their complexity. If so the experimental data represent a starting point for the formulation of new theoretical models to interpret this physical and geophysicist phenomenon.

Earthquakes

1-The first case: the Japan earthquake occurred March 11, 2011. On the occasion of the catastrophic earthquake in Japan (Straser, 2011) gravimeter recorded the event with the low-frequency signals that lend themselves to important interpretations (Fig.5).

The unit was operating with 10 measures about 10 oscillations and, therefore, obtaining average values of the gravity of 15 seconds inserted into intervals of 135 seconds, by Mario Campion' Gravimeter. Under these conditions, the analysis of it is done in some detail, namely, with a total of 650 daily measurements.

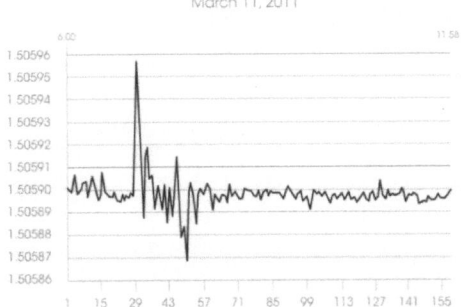

Figure 5. Trend of gravity during the earthquake in Japan on March 11, 2011

The graph shows the registration of an interval of 6 hours and fully grasp the event, with continuous recording from 6 am to 11.58 before noon. The peak on the graph has a value of about 55 microseconds, but we know that it is a value averaged over 15 seconds we do not know if the apparatus has missed its maximum value. In the same event, however, extended over time, with the abscissa a total of 19 measurement intervals (Fig.6). From the graph we see that the phenomenon shows the serious decrease of 40 millionths of a g and is damped with half-wave only.

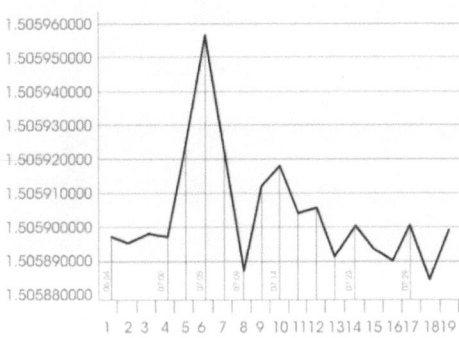

Figure 6. Japanese earthquake in Japan on March 11, 2011

"Be it a phenomenon of the gravitational nature is known from the fact that the whole wave would have persisted for about 13 minutes and then with a frequency certainly not comparable to that of the seismic waves", as commented Dr. Mario Campion. The travel time can be deduced from the data of the INGV, which reported that the quake occurred at 6:46:24 Italian time, while the graph expanded deduce that the same be collected by the gravimetric instrument at 07:05: 00 with a delay of 1241 seconds, traveling at a speed of about 7 km per second.

The graph shows an important detail, and that the symmetry in the variations of gravity, both in correspondence of the peak, both in subsequent surveys which show the variations of gravity that occur according regular times. One feature, the latter, which opens up new questions.

2-A second type is the one to "beak" and, also in this case is respected the temporal regularity of the variation of gravity. This kind of graphics are characterized by a symmetry of the graph that presents, in the vertex, an inclination towards the right. This graphical respected the time interval between the onset of the perturbation and the maximum point, and between this and the negative peak, which is also inclined to the right. In addition, changes in the slope of the change trend of gravity take place at regular intervals. The graph (Fig.7), which shows the increase of gravity in correspondence of the shock, refers to an earthquake of magnitude M7.3 occurred in Russia, and the graph (Fig.8) of an earthquake of M7.3 occurred in Iran whose performance is similar to the preceding quake.

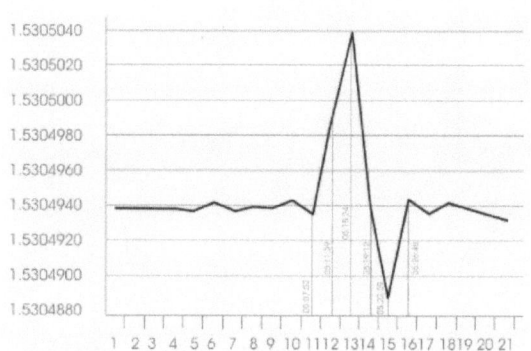

Figure 7. Russian earthquake, sea event on August 14, 2012

Conclusion

We can conclude that, as already noted Maurice Allais, during eclipses occur anomalous trends of gravity, observed in this study even during strong earthquakes.

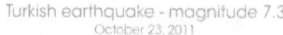
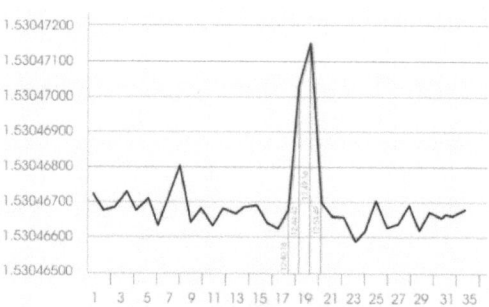

Figure 8. Turkish earthquake, October 23, 2011

The question that remains open is whether to chair the phenomenon is the Newtonian gravitational forces, namely those related to the interaction of the mass of the bodies, or is due to electromagnetic interactions, can influence the movement of the pendulum, and determine the fault gravity. The debate remains open.

Acknowledgements

I would like to express a heartfelt thanks to Dr. Mario Campion for his commitment to scientific progress and for allowing me to analyze the data of his study, and to Prof. Martin Kokus for the encouragement to continue the research that has provided me over the years. Another also thanks to the Company AB Global Service for allowing my participation in this Conference.

References

M.E. Contadakis, and G.Asteriadis, *Natural Hazards and Earth System Sciences*, **1** (2001).

Y Imanishi, T. Sato, T.Higashi, W. Sun and S. Okubo, *Science*, **306**, 5695 (2004).

M.Kokus, Geophysical Research Abstracts, **15**, EGU2013-2679 EGU General Assembly (2013).

V. Straser, *New Concepts in Global Tectonics Newsletter*, **57** (2010).

Straser, V., *New Concepts in Global Tectonics Newsletter*, **59** (2011).

T. Tanaka, W. Salden, A.J. Martin, H. Saegusa, Y. Asai, Y. Fujita and H.Aoki. *Geochemistry Geophysics Geosystems*, **7**, Q03017 (2006)

RELATIVITY AND CONSERVING ENERGY

BOB TICER

3588 Elmira RD
Eugene, Oregon 97402/USA

Relativity theory is shown to comply with conservation of energy in view of a Perfect Cosmological Principle and a Multiverse. It includes a detailed analysis of the Schwartzschild Metric indicating a different interpretation of the Schwartzschild singularity for it to agree with conservation of energy, the equivalence principle, Tired Light, the Mach principle and the Cosmic Coincidence.

Interpreting the Schwartzschild Metric

Conservation of energy is at issue with both special and general relativity insofar as a change in state of the observer questions whether all the mass-energy at large relatively increases or decreases the same. A remedy is here proposed in view of a multiverse that entails analyses of Lorentz transformations and the Schwartzschild Metric.

Lorentz transformations for time t' and distance x' in a direction of relative motion v in relation to light speed c are

$$x' = \frac{x - vt}{\alpha} = \frac{x - vt}{\sqrt{1 - \beta^2}} = \frac{x - vt}{\sqrt{1 - v^2 c^{-2}}} \quad (1)$$

$$t' = \frac{t' - vxc^{-2}}{\alpha} \quad (2)$$

Invariance of the interval s is of the form

$$s^2 = x'^2 - t'^2 = x^2 - t^2 \quad (3)$$

In contrast, the comparable Schwartzschild Metric for the value of v with regard to gravitational potential is of the form

$$ds^2 = c^2 dt^2 \alpha^2 - dx^2 \alpha^{-2} \quad (4)$$

In Eq. (4) the polar coordinates of gravity are disregarded, as were y and z coordinates for relative motion.

Of significance here is the Schwartzschild singularity of the form

$$ds^2 = c^2 dt^2 (0)^2 - dx^2 (0)^{-2} \quad (5)$$

It is assumed a black hole exists at an event horizon as having escape velocity v = c, and the singularity exists within the event horizon in the sense spacetime becomes volume-less as mass-energy becomes infinite.

A main criticism of this result is it violates energy conservation. Jacob Bekenstein pointed out in a 1972 journal that a black hole is inconsistent with the 2nd law of thermodynamics [1]. Stephen Hawking agreed and he modified the black for it to emit Hawking radiation [2], which somehow needs to also apply to the singularity to allow all the mass-energy of the universe to have expanded from it. Moreover, energy conservation does not necessary apply to either relative motion or gravity in the sense that mass-energy at large differs with regard to a relativistic change in local spacetime coordinates as the means of observation.

Interpretation of the singularity is arbitrary in that spacetime and mass-energy in Eq. (5) can apply to either of two terms. Another interpretation of (4) maintains c as relatively constant in a gravitational field as well as with regard to relative motion, as in view of the derivation

$$0 = c^2 dt^2 \alpha^2 - dx^2 \alpha^{-2}$$

$$dx^2 \alpha^{-2} = c^2 dt^2 \alpha^2$$

$$\frac{dx^2}{dt^2} = c^2 \alpha^4$$

$$\frac{dx}{dt} = c\alpha^2 = c \left[1 - \frac{2GM}{xc^2} \right] = c' \quad (6)$$

Consider a local event in a gravitational field. By Eq. (6) light speed is $c' = c\alpha^2$, but by Eq. (4) the radial distance is only $x' = x\alpha^{-1}$. For slower c' it is relatively $x\alpha$. A period of orbit T' is relatively longer as $T\alpha^{-1}$. A clock rate t' is slower as $t\alpha$, such that the slower clock nullifies the shorter distance of slower c. The slower orbital period is also representative of a natural clock.

Consider how the local observer in the field perceives an event outside the field. Due to having a slower clock, light speed is either faster or space-time is relatively greater. We assume the latter. Consider also gravitational potential GM/x is constant in ratio of mass M to radius x. A constant speed for a greater radius thus implies greater mass and energy. The exact opposite applies with regard to an outside observer perceiving an event inside the field. What this implies is mass-energy of gravity is spent in condensing to a smaller volume of mass. However, it still does not explain conservation of mass-energy at large with regard to a local observer's change of state.

Conserving Energy

Consider mass continually absorbs energy of space as a composite of matter and antimatter interacting when condensed together for them to convert into gravitational radiation. The wake of the emitted energy constitutes a vacuum effect as gravity. If the emission of radiation is according to the Doppler principle, then the equivalence of inertial mass and gravitational mass applies.

Consider the emitted radiation is also according to a tired light theory such that it is gradually reabsorbed by spacetime for it to be a relatively weak long range effect in agreement with both the Planck constant h and upper limit c for an escape velocity of a particular radius R_u of the observable universe. Gravitation radiation thus complies with conservation laws of momentum and energy, as gravitational attraction itself is a ripple effect caused by the vacuum effect of emitted radiation.

The tired light principle further provides a recycling process to maintain a constant ratio of gravity to matter. In view of a Cosmic Coincidence the Hubble Constant $H_o = 70.5$ kilometers per second per one million parsecs multiplied by twice the Bohr radius r_a of the hydrogen atom per light speed equals the ratio of gravitational potential to electrostatic potential between two hydrogen atoms each of mass m_a. Replacing electric and magnetic permeability constants of the electrostatic unit

of charge e with H_o, as containing them, the equation becomes

$$\frac{H_o\left(2r_a\right)}{c} \cong \frac{Gm_a^2}{e^2} \cong 8.08 \times 10^{-37} \qquad (7)$$

Replacing H_o with c per radius R_u of the universe, and replacing e^2 with its parameters as $m_a v^2 r_a$, Eq. (7) further equates as

$$\frac{2m_a v^2}{R_u} \cong \frac{Gm_a^2}{r_a^2} \qquad (8)$$

The gravitational force between two touching hydrogen atoms thus approximates to their centripetal force if they rotate from each other from center radius R_u.

Consider now a non coincidence of c as both the escape speed of the observable M_u of the universe and the spin rate of the electron of mass m_e and radius r_e according to the equation

$$\frac{2GM_u}{R_u} = \frac{3e^2}{2m_e r_e} = c^2$$
$$\cong \left[\frac{H_o\left(2r_a\right)e^2}{Gm_a^2}\right]^2 \qquad (9)$$

Eq. (9) suggests M_u and R_u are fixed with regard to H_o and atomic and electrostatic parameters.

To explain how Eq. (7) complies with Eq. (9) the Schwartzschild Metric is reinterpreted. Suppose a radial contraction of spacetime is according to the relativistic factor squared instead of only the relativistic factor. No effect on local clocks thus occur, as slower speed nullifies shorter distance. The Schwartzschild Metric itself is nullified with universal time along with conservation of mass-energy.

To the contrary, according to the principle of equivalence of general relativity, the light spectrum is blue shifted if entering a gravitational field and red shifted if leaving it. The shifts in light spectrum naturally occur as such if clocks in the field are relatively slow.

There is still the possibility of space-time contracting by the relativistic factor squared by combining relativistic effects of relative motion and gravity, as is according to the equivalence principle whereby Einstein equated gravitational free fall with inertial motion in that there is no internal awareness of either one. However, the free fall of Earth with its moon is felt as ocean tides caused by the stronger pull of gravity on parts of Earth closer to the moon. These tidal effects along with the red and blue shifts in light spectrum occur because gravity is inhomogeneous by nature.

The Mach principle can also apply where mass inertia is conditional to its relative distribution. If this relative distribution connects with the equivalence principle and the inhomogeneous nature of gravity, then a homogeneous distribution of mass throughout the universe on a large scale is everywhere the same. Moreover, according to a Perfect Cosmological Principle the universe on a large scale is everywhere the same.

It is estimated from astronomical observation that the scale for homogeneity increases along with greater size. It is thus evident that the universe becomes homogeneous and gravitational massless for relative orbital motion to apply in place of gravitational potential for a contraction according to the relativistic factor squared in consistency with light speed also slowed by the relativistic factor squared. In effect, the Lorentz Metric is simply nullified on the larger scale.

With tired light applying to all light, the red shift in light spectrum reverses to the outer edge of the universe whereby the observer is at center. In effect, the universe becomes an observable part of an infinite universe that is essentially a multiverse.

1. Implications

In the early days of big bang theory (1970s) it was suggested by a professor that it would provide a frame of reference for observational study even if it were not true. The multiverse provides a more inclusive frame of reference. It even includes big bang theory inasmuch as it is still possible the observable part of the universe that we are in is expanding. Quantum Electrodynamics allows it according to probability. Small events of least energy occur most often, whereas expansion of an entire universe occurs only once in about 13.8 billion years. In view of the multiverse, this expansion is allowed by its change in what is observed.

There are countless ways of describing and explaining phenomena. A useful frame of reference is to include as many ways as are possible. For instance, a plenum could apply with regard to matter and antimatter being according to superposing opposite spin momenta. Since a plenum allows no variation in density of actual substance, mass and volume ratios as calculated by Carl R. Littmann could apply [3]. In email correspondence Greg Volk pointed out a number ratio of 9.89898 according to a tetrahedral pattern of 4 spheres packed around a central one, and Harold Aspden also related $(9.89898 - 1)$ to the ratio of the proton and electron masses according to a combination of total energy of Charles-Augustin de Coulomb's formula for electromagnetic attraction and Joseph John Thomson's formula for internal mass energy. Similarly, Charles Lucas Jr. combined the Lorentz force with Ampere's law of induction in arriving at a 4th order effect of gravity, as a drift velocity, in comparison to the force of electromagnetism [4]. More detail of this development is provided in a free pdf of my book A Mystic History In Light Of Physics: http://bobticer.com.

References

1. J. D. Bekenstein, Black Holes and Entropy, *Phy. Rev.* **D7 (8),** 2333-2346 (1975).
2. S. W, Hawking, Particle Creation by Black Holes. *Cosmic Math Phy.* **43,** (1975) 199-220 (1975).
3. C. R. Littman, Sphere Volume Ratios in Tetrahedral and Triangular Patterns, and Some Implications, *NPA* **18,** 354-318, (2011).
4. C. W. Lucas, Electrodynamic Origins of Gravitational Forces & 'The Universal Electrodynamic Force. *NPA* **18**. 375-397, (201

ASSUMPTIONS CONCERNING THE INTERPRETATION
OF THE MICHELSON-MORELY EXPERIMENT

ALEKSANDR TSYBIN

Haldeman Ave, Apt 71A
Philadelphia 19115, PA
Email: acibin@yahoo.com

The famous Michelson-Morley experiment yielded negative results. To justify this result, Einstein formulated his famous postulates. This article is one explanation of the different outcome of this experiment, based on the well-known formula of Planck.

Einstein's postulate states that the speed of light c- is constant and does not depend on the direction and velocity of the source. This postulate provides an opportunity to convincingly justify the negative results of the Michelson-Morley experiment and abandon the concept of the cosmic ether. But in present time, a whole series of facts have arisen which give opportunity to doubt such an interpretation of this phenomenon, in particular that space consists solely of emptiness [1,2,3, 4]. The facts included are: that space has a temperature, that in space, there are eddies, convective cells and solitons (solitary wave), and finally that, in space, Casimir's effect has been observed [5,6]. Not merely by chance that space, which many authors call the physical vacuum, is not a vacuum or a void in the truest sense of the word, but continuous. And yet, attempts to conclusively prove a negative result Michelson-Morley experiment have not succeeded. What is more, attempts to repeat this experiment have led to the same negative result [7,8,9]. Let us briefly recall that significant negative results will include the difference between two different time course of the light beam in the initial position and when the interferometer is turned by 90^0 and by $v \ll c$ on was to be

$$\Delta t = \Delta t^{(1)} - \Delta t^{(2)} =$$
$$= \frac{l_1 + l_2}{c} \frac{v^2}{c^2} \quad (1)$$

Here $\Delta t^{(1)}$ – is the difference in time course of the light beam in the initial position of the interferometer and $\Delta t^{(2)}$ –is the difference in time course of the light beam after the interferometer has been rotated 90^0. l_1 the company and l_2, accordingly, are the horizontal and vertical shoulders interferometer. v – is the velocity of the Earth's orbit around the Sun. Time is measured in seconds, the length of the shoulders in meters, and the velocity meter/sec value is obtained as the total length

of 30 meters apart, but the light for a time travels a distance equal to 3×10^{-7} meters and the distance it is easy was to measure, while the experiment showed the distance is exactly equal to zero. This is a negative result.

The situation seems to me to be reminiscent of what happened with the known dependence of the Rayleigh-Jeans Law [10, 11] describing the radiation of black bodies. According to this dependence, the energy density of blackbody radiation is equal to:

$$U_\omega = \frac{\omega^2}{2\pi^2 c^3} <E>, \quad (2)$$

where U_ω is the energy density of blackbody radiation in the $j \times sec/meter^3$, j – in the Newton $\times$ meter, and ω – is the angular frequency. 1/sec in this case refers to the angular frequency of oscillation of the light wave. $<E>$ – shows the average energy fluctuations in the single oscillator j. It is well known that the energy of a one dimensional oscillator is expressed in terms of p in the kg $\times$ meter/sec and coordinate q in the meters as follows:

$$E(p,q) = \frac{p^2}{2m} + \frac{m\omega^2 q^2}{2}, \quad (3)$$

where m – is the mass in kilograms. In classical statistics, the equilibrium distribution of particles (in this case the oscillator) is calculated by the formula

$$W(E) = A \times \exp(-E/T), \quad (4)$$

where A – represents the positive constant and T-absolute temperature in degrees Kelvin. Therefore, the average energy is

$$<E> = \frac{A \iint E(p,q) \exp(-E(p,q)/T) dpdq}{A \iint \exp(-E(p,q)/T) dpdq}. \quad (5)$$

We introduce the notation $P = \frac{p}{\sqrt{2m}}$, $Q = q\,\omega\sqrt{\frac{m}{2}}$ and

then
$$<E> = \frac{\int P^2 \exp(-P^2/T)dP}{\int \exp(-P^2/T)dP} +$$

$$+ \frac{\int Q^2 \exp(-Q^2/T)\,dQ}{\int \exp(-Q^2/T)\,dQ},$$

That is, it reduces it to the computation of two integrals:

$I_0 = \int_{-\infty}^{+\infty} \exp\left(-\frac{x^2}{T}\right) dx$ and $I_2 =$
$\int_{-\infty}^{+\infty} x^2 \exp\left(-\frac{x^2}{T}\right) dx$. which are converted by $y = \frac{x}{\sqrt{T}}$ replacing the integrals:

$$I_0 = \sqrt{T} \int_{-\infty}^{+\infty} \exp(-y^2)\,dy$$

$$I_2 = T\sqrt{T} \int_{-\infty}^{+\infty} y^2 \exp(-y^2)\,dy.$$

The second integral can be expressed through the first through the integration by parts:
$\int_{-\infty}^{+\infty} y^2 \exp(-y^2)dy = \frac{1}{2}[-y\exp(-y^2)I_{-\infty}^{\infty} +$
$+ \int_{-\infty}^{+\infty} \exp(-y^2)dy] = \frac{1}{2}\int_{-\infty}^{+\infty} \exp(-y^2)dy.$

And we will eventually come to the well-known Gaussian integral [12]. It can be easy to calculate
$\int_{-\infty}^{+\infty} \exp(-z^2)dz =$

$$\sqrt{\int_{-\infty}^{+\infty} \exp(-x^2)dx \int_{-\infty}^{+\infty} \exp(-y^2)dy} =$$

$$\sqrt{\int_{-\infty}^{+\infty}\int_{-\infty}^{+\infty} \exp(-(x^2+y^2))dxdy}.$$

And then we pass to polar coordinates $x = r\cos(\phi)$, $y = r\sin(\phi)$. As a result, the last double integral is transformed to:

$\int_0^{2\pi} d\phi \int_0^{\infty} \exp(-r^2)rdr = -2\pi\frac{1}{2}\exp(-r^2)I_0^{\infty} = \pi.$

Hence: $I_0 = \sqrt{T}\sqrt{\pi}$, $I_2 = T\sqrt{T}\frac{1}{2}\sqrt{\pi}$.

And then:

$$<E> = T. \tag{6}$$

And so the formula (2) takes the form:

$$U_\omega = \frac{\omega^2}{2\pi^2 c^3} T. \tag{7}$$

However, formula (7), as is well known, does not reflect the physical picture of the phenomenon and cannot be improved upon by using methods of classical physics. This is where the Max Planck hypothesis comes to our aid, explaining that the emission and energy absorption do not occur continuously, and giving the portions of energy (Photons). Sam oscillator is in discrete energy states:

$$E_n = n\hbar\omega, \tag{8}$$

where $\hbar$ is the modified Planck's $\hbar = \frac{h}{2\pi}$ constant and h is Planck's constant. Then, according to Planck's hypothesis, the average energy of the one-dimensional oscillator is expressed by the relation:

$$<E> = \frac{\sum_{n=0}^{\infty} E_n \exp\left(-\frac{E_n}{T}\right)}{\sum_{n=0}^{\infty} \exp\left(-\frac{E_n}{T}\right)}$$

and if we introduce the notation $x = \frac{\hbar\omega}{T}$ than the latter designation can be rewritten as:

$$<E> = \hbar\omega \frac{\sum_{n=0}^{\infty} n\exp(-nx)}{\sum_{n=0}^{\infty} \exp(-nx)}.$$

It is easy to see that the denominator of this expression is the infinitely decreasing geometric progression with denominator $\exp(-x)$ and therefore it is equal to the sum $\frac{1}{1-\exp(-x)}$ in the numerator and the derivative of this amount x of that $\frac{\exp(-x)}{(1-\exp(-x))^2}$ has the result that we finally obtain

$$<E> = \frac{\hbar\omega}{\exp(\frac{\hbar\omega}{T})-1}. \tag{9}$$

This formula is well known and represents a quantum analogue of Eq. (6). Instead of the formula we obtain the Rayleigh-Jeans formula Plank.

$$U_\omega = \frac{\omega^2}{2\pi^2 c^3} \frac{\hbar\omega}{\exp(\frac{\hbar\omega}{T})-1}. \tag{10}$$

As is well known in the classical case

$$\hbar \to 0 \; \exp\left(\frac{\hbar\omega}{T}\right) \to 1 + \frac{\hbar\omega}{T}$$

and expression (10) becomes (2), as expected. Similarly, the classical expression (1) can be represented in the quantum representation, that is, instead of (1) should be written next phenomenological formula:

$$\Delta t = \Delta t^{(1)} - \Delta t^{(2)} = \frac{l_1 + l_2}{c} \frac{v^2}{c^2} \times D(\hbar, \omega), \qquad (11)$$

$$D(\hbar, \omega) = \frac{-4T^2}{\pi \hbar^2 \omega^2} \times \cos\left\{\frac{\pi}{2}\left[\exp(\frac{\hbar \omega}{T}) - \frac{\hbar \omega}{<E>}\right]\right\} = 0,$$

which in quantum representation is the case when $\hbar > 0$ is the expression in square brackets. Accordingly (9) is identically equal to unity. In the classical case $\hbar \to 0$:

$$D(\hbar, \omega) = \frac{-4T^2}{\pi \hbar^2 \omega^2} \times \cos\left\{\frac{\pi}{2}\left[1 + \frac{\hbar \omega}{T} + \frac{\hbar^2 \omega^2}{2T^2} - \frac{\hbar \omega}{T}\right]\right\} \to$$

$$-\frac{4T^2}{\pi \hbar^2 \omega^2} \times \left\{\cos(\frac{\pi}{2})\cos(\frac{\pi \hbar^2 \omega^2}{4T^2}) - \sin\left(\frac{\pi}{2}\right)\sin\left(\frac{\pi \hbar^2 \omega^2}{4T^2}\right)\right\} \to$$

$$\left(-\frac{4T^2}{\pi \hbar^2 \omega^2}\right) \times \left(-\frac{\pi \hbar^2 \omega^2}{4T^2}\right) = 1. \quad (12)$$

This is the classic version of c taking into account formula (6), expression (11) leads, as expected, the relation (1). It follows that formula (11) qualitatively correctly reflects the existing pattern phenomenon, especially as the sun from a physical point of view, a black body and a constant $\frac{\pi}{2}$ plays an important role in construction of the Michelson-Morley interferometer. And so the indirect connection between the formulas (10) and (11) can be justified. The truth is justified in any case cannot be considered proof of the correctness of Eq. (11) and it is only an analytical dependence phenomenological correctly reflects the physical picture of the phenomenon. Similar is a phenomenological dependence of Wien's law. We only note once again that in the case of the validity of (11) need to postulate that there is no Einstein.

P.S. Incidentally, the function $y(\tilde{x}) = \cos\left\{\left[\exp(\frac{\hbar \omega}{T}) - \frac{\hbar \omega}{<E>}\right]\tilde{x}\right\}$ satisfies the equation $\frac{d^2 y}{d\tilde{x}^2} + y = 0$ and the homogeneous boundary conditions of $\frac{dy}{d\tilde{x}}(0) = y\left(\frac{\pi}{2}\right) = 0$. It is one of the homogeneous functions of the boundary problem. Here $\tilde{x}$ —the dimension-less spatial coordinate. This equation can come from a known one-dimensional equation. With respect to the Schrödinger wave function $\psi(\tilde{x}, t)$

$$\frac{\partial}{\partial t}\psi = \frac{-\hbar}{2mR^2}\frac{d^2 \psi}{d\tilde{x}^2}$$

R —which defines the size in meters and t —time in seconds. Then do the following

$$\psi(\tilde{x}, t) = \exp\left(-i\frac{\hbar t}{2mR^2}\right) \times y(\tilde{x})$$

change.

Anyway. Perhaps the following argument.

$$|R\exp(i\alpha)|^2 + |R\exp(i\beta)|^2 -$$

$$-|R\exp(i\alpha) + R\exp(i\beta)|^2 =$$

$$= R^2 + R^2 - 2R^2[1 + \cos(\alpha - \beta)] =$$

$$= -2R^2\cos(\alpha - \beta).$$

Where $R = \frac{\sqrt{2}T}{\sqrt{\pi}\hbar\omega}$ and $\alpha = \frac{\pi}{2}\exp(\frac{\hbar \omega}{T}), \beta = \frac{\pi}{2}\frac{\hbar \omega}{<E>}$.

Then we have

$$Y = -4\frac{T^2}{\pi \hbar^2 \omega^2} \times \cos\left\{\frac{\pi}{2}\left[\exp(\frac{\hbar \omega}{T}) - \frac{\hbar \omega}{<E>}\right]\right\},$$

but accordingly (9) will be $\exp(\frac{\hbar \omega}{T}) - \frac{\hbar \omega}{<E>} = 1..$

We find from this that Y=0.

Conclusion

The article contains a formula that correctly describes the phenomenological results of the Michelson-Morley experiment. In the classical case, where the Planck's constant is assumed to be zero, it gives a well-known negative result. In the case where Planck's constant is not zero, this formula is identically zero. If these arguments are valid, then there is no need for Einstein's postulate.

References

1. A.V. Rykov. Vacuum and Substance of the Universe (M, 2007) 289 pp.
2. A. Tsybin: Another Deduction of Einstein's Formula: Proceedings of the Natural Philosophy Alliance. 16th Annual Conference of the NPA, Vol. 6.No.2. pp. 298-299.
3. A. Tsybin: On the Increase of Particle Mass with Velocity. Proceedings of the Natural Philosophy

Alliance. 17- th Annual Conference of the NPA, Vol. 7. pp. 582-582.

4. A. Tsybin: Space vs. Vacuum: Facts Show That Space Isn't a Vacuum. Proceedings of Natural Philosophy Alliance. 18th Annual Conference of NPA, Vol. Eight. pp. 646-650.

5. James F. Babb, V. Hushwater, "Casimir Effect Bibliography" Am. J. Phys. 65 (5): 381-384 (May 1997).

6. Dzyaloshinskii I.E.; Kats E.I. (2004) "Casimir forces in modulated systems". Journal of Physics: Condensed Matter 16 (32): 5659. ArXiv: condmat / 0408348. Bibcode 2004 JPCM ... 16.5659D. Doi: 10.1088 / 0953-984/16/32/003.

7. Conference on the Michelson-Morley experiment. Held at the Mount Wilson Observatory, Pasadena, California, February 4 and 5, 1927) / / The Astrophysical Journal. December 1928.Vol. 68, No. 5.P. 341-402.

8. R.S. Shankland, S.W. McCuskey, F.C. Leone, and G. Kuerti. New Analysis of the Interferometer Observations of Dayton C. Miller / / Rev. Mod. Phys 1955. - Vol. 27. -P. 167-178-DOI: 10.1103/RevModPhys.27.167.

9. Swenson.L.S. The Michelson -Morley-Miller Experiments before and after 1905 / / Journal for the History of Astronomy-1970-Vol. – P.56-78.

10. Max Planck. An improvement in radiation law Wines. Favorites works. Science. Moscow. In 1970. page 249.

11. Max Planck. The theory of normal distribution of the radiation energy spectrum. Selected Works. Science. Moscow. In 1970. page 251.

12. I. S. Gradstein & I. M. Ryzik. Tables of Integrals, Sums, Numbers and Products. p. 1108 (The State Publishing House, Physics- Mathematics Literatures, Moscow, 1963). Russian Language.

ARE THERE BLACK HOLES?

ALEKSANDR TSYBIN

9926 Haldeman Ave,, Apt 71A
Philadelphia, 19115, PA, USA

Email: acibin@yahoo.com

The possible existence of black holes have been discussed in scientific literature under various aspects. The prevailing view is that such an objects do exist, and have a tremendous gravitational attraction. The escape velocity of these objects would exceed the speed of light in vacuum, and thus even light cannot escape. But there are arguments against the existence of black holes, one of which is presented below. As a consequence, an alternative interpretation of gravitation and the physical vacuum is suggested.

Ref. [1] gives relative change in wavelength of light (red shift), emanating from a cosmic body of weight M_t [kg] (Earth) and of radius R_t [m] Earth as

$$\frac{\Delta\lambda}{\lambda} = \gamma_c \frac{M_t}{R_t c^2}\sqrt{\frac{r}{R_t}}\ln\left(\frac{r}{R_t}\right), \quad r \geq R_t, \quad (1)$$

where r[m] is the distance from Earth, $c = 3 \times 10^8$ meter/sec is the vacuum velocity of light, and γ_c —is the constant of gravity $\gamma_c = 6.67 \times 10^{-11}$ meter^3/kgsec^2.

As shown in Ref. [1], this formula qualitatively renders the observations taken from Earth for large and small r. The formula (1) can be rewritten as:

$$\frac{\Delta\lambda}{\lambda} = \frac{v_t^2}{2c^2}\sqrt{\frac{r}{R_t}}\ln\left(\frac{r}{R_t}\right), \quad r \geq R_t, \quad (2)$$

where $v_t = \sqrt{2g_t R_t}$ —is the escape velocity for this cosmic body with $g_t = \gamma_c \frac{M_t}{R_t^2}$ being the acceleration due to gravity at the surface of the body [$\frac{\text{meter}}{\text{sec}^2}$.] If this object is a black hole, then we get

$$\frac{\Delta\lambda}{\lambda} > \frac{1}{2}\sqrt{\frac{r}{R_t}}\ln\left(\frac{r}{R_t}\right), \quad r \geq R_t, \quad (3)$$

which is a too large of a value. For example, if the center of our galaxy has a black hole with a radius less than one light year, the distance from the Earth to the center of the galaxy is about 25,000 light years, and formula (3) yields an approximate red shift equal to $\frac{\Delta\lambda}{\lambda} > 16$.

Such large values of the red shift have never been observed anywhere. We get a paradox which suggests that black holes don't exist. These ideas were expressed earlier [2,3], but they are rejected by official astronomy because the black hole is still one of the most popular objects in astrophysics. However, an article published on January 22, 2014 [4], argues that no black holes exist. This is sensational, since its author, S.W. Hawking, is considered one of the most authoritative experts on the theory of black holes. Hawking came to this conclusion on the basis of the so-called information paradox to which dozens of articles were devoted [5, 6]. What object might then be located, from our point of view, in the center of the spiral galaxies?

We begin with a summary of the problem with gravity [7,8]. Newton's gravitational force F between two bodies reads $F = \gamma_c \frac{Mm}{r^2}$, where F is measured in N (Newton's), M and m – mass of the two bodies are in (kg), and the distance r between the centers of gravity of the two masses is in meters. Rewriting the formula as $w = \frac{Fr^2}{m} = \gamma_c M$, gives the quantity of w as having the dimension of $\left(\frac{\text{meter}^3}{\text{sec}^2}\right)$. This dimension indicates the change of a vector flux. We may conclude from the inspection of the dimensions and its vector properties that the distinction is as important as the very well-known example demonstrates: the dimension has a (Newton × meter) scalar meaning work and a vector meaning torque. Here we assume as our hypothesis that a vector flux vector of dark matter and possibly dark energy fill the so-called physical vacuum. In the early 20th century, after Einstein's papers, it was accepted that there is no ether, that we are surrounded by a vacuum. I remember that the question: "what is the temperature of the vacuum?"

was replied with: "the vacuum is a void that has not and cannot have any temperature." It is now generally accepted that the so-called "void" does have a temperature. This temperature was measured and is equal to 2.725^0,K (a temperature of the maximum of the background radiation.)

The so-called "vacuum" has spiral, elliptical and irregular vortices. We live inside one of these spiral vortices. Spiral vortices are well known in the atmosphere and hydrosphere of the Earth. These and other types of vortices have also been observed in the atmospheres of Jupiter, Saturn and the Sun. Recently, Mark Vogelsberger and his staff have per-performed a numerical computer experiment using the Harvard supercomputer "Odyssey" with the latest software Arepo. The computer worked continuously for several months and eventually they observed spiral vortices arsing in a virtual dark matter. The time spent on the computer corresponds to about 14 billion years old in reality, and virtual elliptical vortices have not appeared during the specified time. It is now believed that about three billion years from now our Milky Way galaxy will unite with the neighboring Andromeda galaxy, and together they will form an elliptical galaxy. The Sun, (one of the class of stars that live to become about 7.8 billion years old) or what will remain of it, will be at a distance of 160 thousand light years from the center of the new elliptical galaxy.

The so-called "vacuum" consists of convective cells of the hexagonal honeycomb type. The linear size of these cells is about 200 Mpc. Similar cells, much smaller but exceeding the diameter of the Earth, have been detected in the atmospheres of Saturn, Jupiter and the Sun. This is more evidence that the so-called "vacuum" is not just an empty void.

In the so-called "vacuum", solitons (solitary wave) were discovered, much like those previously observed in a gaseous or liquid form in an optical medium which are tangible materials, rather than a vacuum.

In the so-called "vacuum" the Casimir effect takes place, analogously it also exists in a gaseous and a liquid medium.

In the so-called "vacuum" there is real emptiness (voids) in deep space, ranging in size from 100 Mpc to 3000 Mpc [9], where there are no galaxies and no stars, and hence no planets. Analogues of this void phenomenon are air holes in atmosphere and gas bubbles in a liquid.

In the so-called "vacuum" there is effect of the added mass, due to which the body weight increases. It takes place in hydrodynamics, aerodynamics and in this "vacuum."

The above evidence supports the fact that the "vacuum" in space is not a void, but a peculiar medium having physical characteristic such as a flux. Now the law of gravity can be interpreted as follows: **the rate of change of the flux passing through a cosmic body is directly proportional to the mass, and the coefficient of proportionality is the constant of gravity.** This means that there will be only attraction between two masses and never repulsion. Incidentally, according to the statements of his contemporaries, in an attempt to answer the question what makes bodies attract one another, Newton replied something like this. "The more I think about it, the more it seems to me that we live in some kind of liquid that flows through us quietly." This view was developed further by Fatio in the early 18th century [10]. In response to criticism that small liquid particles will bounce off of the surface of a body, he predicted that for this hypothetical liquid all surrounding bodies are porous and, even more, consist mainly of empty space. Of course, he could not allow such a heretical idea that an atom of any substance is virtually empty and particles such as neutrinos can penetrate through it unscathed.

Recently, experiments failed to detect a different flavor of neutrinos, which were called sterile neutrinos[21]. They were named sterile because they are not even involved in the weakest interactions. Sterile neutrinos, with significant mass (its mass must be higher than that of the tau neutrino) may become one of the main candidates for the dark matter particles in contrast to the hypothetical WIMPs, which have not yet been detected. Fatio's hypothesis was developed subsequently by Le Sage[11]. J. J. Thomson [12] considered the model of Le Sage and suggested that the flow of extraterrestrial particles is a hypothetical form of radiation. With much is more penetrating than X-rays. Neutrinos have this property. It is worthwhile to mention the work of Henry Cavendish [13], Ettore Majorana [14] in the past and, at the present, work by Robert de Hilster [15,16], as well as an extensive review of S.G.Fedosin [17]. Neutrinos cannot penetrate the nucleus and therefore, according to the gravity hypothesis presented here, there is no gravitational interaction between the atomic nucleus and the electrons. Typically gravitational interaction is neglected as compared with the electromagnetic interaction, there being an enormous difference between these forces. Here the difference is not considered insignificant, but it exactly equals zero. The same argument applies, from our point of view, to neutron stars and so-called black holes. Neutrinos pass through the body, the density of which is $\rho \geq 1022\frac{gramm}{sm^3}$ and therefore, in the vicinity of their surface, no gravity occurs. These bodies represent the analog of an impermeable screen for so-called physical vacuum. This means that in the center of spiral galaxies, including the Milky Way, there is no such body which exists due to the huge gravitational absorption of approaching stars. Instead, this body does not have any power of gravity at all. An indirect confirmation of this conclusion comes from the following astronomical observations [18]. The Hubble Space Telescope observations report a ring consisting of more than four hundred young stars that orbit around a super massive black hole in the center of spiral galaxy the Andromeda nebula, and according to the theory of black holes in principle it should not be there. The stars located in the form of a disk with diameter of one light year, would inevitably have been bro-ken by the gravitational field of the central black hole, not even reaching their embryonic phase of development. Meanwhile, according to the data on board the Hubble spectrograph STIS (Space Telescope Imaging Spectrograph), these stars have been well developed for at least 200 million years. Curiously, the New Scientist SPACE reports that the ring of blue young stars is surrounded by another ring of red giants, extending approximately 5 light years, and lying in the same plane. In some respects, which are not similar, a ring surrounds the central black hole in the Milky Way. For example, those detected in M31 (Andromeda) phenomenon are not so unique and does not fit into the concept of black holes. According to

reference [19], astronomers at the Keck Observatory telescopes discovered the star SO-102, located at a distance of 120 astronomic units. This is comparable to the radius of the solar system, and is negligible when compared with interstellar distances. Despite such close proximity, the star was not absorbed by the black hole, but orbits around it with a velocity of about 10,000 km / sec.

The above observations cannot be reconciled with the standard theory of the existence of black holes, but the assumption of gravity being caused by a neutrino flux can account for it. However, to prove this hypothesis, an experiment is needed to determine the rate of change of the neutrino flux at its passage through the Earth. A neutrino flux was measured in the experiment mentioned in reference [20]. Unfortunately, due to a technical error, this experiment was declared invalid. The repetition of a similar experiment has been postponed at least for a year.

(Additionally, in September 2014, Ref. [21] was published which demonstrates that black hole do not exist.)

Conclusion

Based on some theoretical hypotheses, including works by the present author, as well as on a number of astronomical observations, it is concluded that in the center of spiral galaxies there is a massive object that has no gravity. The situation in the center of these galaxies resembles that inside the eye of a typhoon or hurricane where all is quiet and calm.

References

1. A. M. Tsybin. *Fundamental problems of science and technology.* Issue 35. Part 2. Sankt-Petersburg 2012, 367-369. (in Russian).
2. A.A. Logunov. Lectures on the Theory of Relativity and Gravitation. Modern and analysis of problems. Moskau. "Science." 1987. Russian Language.
3. Jaroslaw G. Klyushin. Fundamental problems in Electrodynamics and Gravidynamcs. Published by Galilean Electrodynamics, Arlington Massachusetts. 2009.
4. S.W .Hawking. Information Preservation and Weather Forecasting for Black Holes. ArXiv . 401. 5761 v1 [help-th]. 22 Jan 2014.

5. A.Almheri, D. Marolf, J. Polchinski, J. Sully, Black Holes: Complementarity or Firewalls . J. Hight Energy Phys.2, 062 (2013).

6. S.W. Hawking. Information Loss in Black Holes, Phys. Rev. D72 084013 (2005).

7. A. M. Tsybin. Hypothesis about universal gravitation . Fundamental problem of Science and technology. A series of problems in the study of the universe. Issue 32 2007.pp. 437-441. Russian Language.

8. Aleksandr Tsybin. Gravity from neutrino flux. Proceeding of the Natural Philosophy Alliance.14- th Annual Conference of the NPA 21-25 May 2007 at the University Connecticut at Storrs ISSN 1555-4775 Vol 4. No 2, pp. 255-257

9. A.V.Rykov. Vacuum and matter of the Universe. (Moskau. 2007), 289 p. Russian Language.

10. Fatio (1701), pp. 32-35; Secondary links , Zehe (1980), pp. 206-214.

11. Le Sage G.L. (1756). «Letter a une academician de Dijon." , Mercure de France; 153-171.

12. Thomson, W. (Lord Kelvin) (1873), "On the ultra mundane corpuscles corpuscle of Le Sage". Phil. Mag. 4 th ser. 45,. 321-332.

13. Brush Stephen G; Holton, Gerald James Physics, the human adventure: from Copernicus to Einstein and beyond – New Brunswick. N. J. Rutgers University Press; 2001, p. 137- ISBN 0-8135- 2908 -5.

14. Ettore Majorana: scientific papers by Bassani, Giuseppe – Franco Societa Italia Di Fisica (paperback). Robert de Hilster , "The Gravitation Equation", presented at NPA -15 (Albuquerque, NM, April 2008).

15. Robert de Hilster, "Majorana's Experiments and the New Equation for Gravity", Presented at the NPA-16 (Storrs, CT, May, 2009).

16. S.G. Fedosin. Physical theory and infinite nesting of matter. Perm, 2009, 844 p., Tabl. 21. Fig. 41, Ref. 289. ISBN 978-5 -9901951-1-0. Russian Language.

17. Viktor Karev : 2:5090 /69. 111, http: // elementy. Ru /images/ news/ Andromeda_rus_300.jpg)

18. www. lenta. ru/ news/2012/10/05/ close-star hole.

19. Erwin Cartlidge, "OPERA Confirms and Submits Results, But Unease Remains" Science Insider (17 Nov 2011), http:// news, science, ag. Org / science insider/ 2011/11/ Faster-than- light-neutrinos-opera .html?ref=hp.

20. Aguilar- Arevalo et al. Event Excess in Oscillations. (http://dx.doi.org /10. 1103/ Rev RevLett.105.181801// Phys Rev. Lett.- 2010- T. 105-#18.

www.ingramcontent.com/pod-product-compliance
Lightning Source LLC
Chambersburg PA
CBHW050716180526
45159CB00003B/1042

In course of time, the influence of north and central India began to grow in the sculptural art of Bangladesh and the introduction of using stones started. From the early three centuries of the common era, the local sculptors started to make black stone sculptures in the Kusana style, native to northern India. These sculptures were the images of the deities worshiped by the followers of the three major religions of the time, namely, Brahmanism, Buddhism andJainism.

Bronze sculptures began to be assimilated in the 7th century AD primarily from the Chittagong region. The earliest sculptures of this kind were depictions of Buddhist believes but the art was later integrated into the Hindu art as well.

In modern times, the theme of sculptural art has been dominated by some historical events, mainly the Bangladesh liberation war. Aparajeyo Bangla, Shabash Bangladesh are some of the noteworthy examples of this trend.

As in other countries of the world, the people of rustic, and primitive ideas developed folk art in Bangladesh. Because of this the structure and growth of the folk-art of Bangladesh are filled with pure and simple vigor and the symbolic representations of hope, aspiration and sense of beauty of the rural Bangladeshi folk. The environment and the agricultural activities greatly helped to enrich the traditional folk-art of Bangladesh. It uses traditional motifs reflecting the land and its people. Different forms of folk art tend to repeat these common motifs. For instance, the lotus, the sun, the tree-of-life, flowery creepers etc. are seen in paintings, embroidery, weaving, carving and engraving. Other common motifs are fish, elephant, horse, peacock, swastika, circle, waves, temple, mosque etc. Many of these motifs have symbolical meanings. For example, the fish represents fertility, the sheaf of paddy prosperity, the lotus purity and the Swastika good fortune. Another factor, most important perhaps, that has influenced the art and culture of this land is the six seasons.

The folk art of Bangladesh has been largely contributed by the rural women because of the aesthetic value as well as the quality of their work. A key reason behind it was that in most cases their art has been non-commercial, whereas the folk art produced by men has a commercial value attached to it. Thus, artists like blacksmiths, potters, cobblers, painters, goldsmiths, brass-smiths, weavers earn their livelihood from what they produce while traditionally, from the past, Alpana artists or Nakshi kantha needlewomen were working within the home and received no monetary recompense for their labor. Both Alpana and Nakshi kantha are some of the most attractive forms of Bangladeshi folk art. Pottery and Ivory are also some popular forms of the art.The movement of modern art in Bangladesh has its roots in the early 20th century. Back then there was no training or educational institutions for arts in Bangladesh. In the late 19th century, the British started to establish some art schools in Calcutta the then provincial capital of Bengal which inspired the local art admirers to pursue a particular form of art. The art lovers of Bangladesh or erstwhile East Bengal were also induced by this. This phenomenon gave birth to many preeminent figures of arts in Bangladesh whose fame spread all the way through not only in Bangladesh but in the whole world. Zainul

Abedin was from this generation of artists. He is considered as the pioneer of art movement in Bangladesh.

After the partition of India, Calcutta became a part of West Bengal in India while the current geographical area of Bangladesh formed the East Pakistan province of Pakistan. Hence, the local artists felt a dire need of an art institution in Bangladesh. In 1948, Zainul Abedin, along with other leading local artists like Quamrul Hassan, Safiuddin Ahmed, Anwarul Huq, Khawaja Shafique established the Dhaka Art Institute to evolve the art tradition in Bangladesh.

Since the establishment of the art institute, the artists in Bangladesh started to gain the much required professionalism and also started to attach commercial value to it. This prompted them to organize art exhibitions to showcase their work to the audiences. By the 1960s the artists started to link with the art traditions of other parts of the world which gained them a pretty clear understanding of contemporary art in those countries. Many artists went to Europe and Japan for training and came back with new ideas and latest techniques, but they were also steeped in the traditions of indigenous art forms.

Bangladeshi potters have been producing exquisite pottery products for ages. Pottery in this region can be traced back to around 1500 BC. Six types of earthenware of high quality have been found in archaeological sites like Mahasthangarh, Govinda Vita, Vasu Bihara, Wari-Batesh-war, Mainamati and Paharpur.

Over the years, clay pottery in Bangladesh has faded but recently it has regained its stature. In the capital itself there are over 700 pottery shops. The traditional potters are called as kumars or khumbas. Kumar or potter family are found all over Bangladesh. The elaborate terra cotta tile works display enormous sweep and dedicated Kumars.

In some communities of kumars make clay pots, vessel for cooking, storing water etc. Other sub-castes fashion figures in the shape of animals, birds, humans and children toys. The "Sakher Hari", an earthen pot, painted with images of fish, combs, birds and floral creepers to denote fertility is used to carry sweets for a marriage ceremony. Whether it is making the unpolished earthenware or glazed ceramics, the traditional potters put in a lot of effort and hard work to put forward their best. The clay pottery is made on a wheel; the potter puts kneaded clay in the center of the wheel and spins the wheel while starting to give the clay a shape simultaneously.

There are three types of pottery namely earthenware, stoneware and a combination of feldspathic material with stoneware. Earthenware is the oldest type of pottery and for ages the techniques of making it hasn't changed. However, with modernity seeping into every aspect of life, the materials and a few alterations in the method have been accepted. Earthenware is made from blending of clays that is later baked hard in fire. After glazing was invented, earthenware was coated with glaze to improve its aesthetic appearance and also to make it water-proof.

The demand for traditional clay pottery from Bangladesh in on a rise and today there are a number of ceramic industries that are functioning in the country.

People visiting Bangladesh are flabbergasted by the sheer beauty and variety of pottery that they can see in shops

Villages and rivers are the backbone of Bangladesh. There are about 68,000 villages in Bangladesh. Most of the people still live in Bangladesh village. Life here is slow, air and food is fresh as opposed to the cities. Food is made from the fresh vegetables directly coming from the field. People make different types of rice-cakes so tasty and authentic.

Profession of most of the people in the villages is farming. But people also have other professions. There are normally different areas in every village based on their profession like potter, fisherman, blacksmith etc. In this photo blog we'll take you to visit life and nature of a remote Bangladesh village locate at southwest part of the country so that you can get an idea about how they are and how people live there.

Preserve tradition of pottery from extinction

Clay potter from Madavpur From the very beginning of our Banglee culture, pottery has represented our identity and lifestyle. The artisans' works include making clay-pots, earthen ware, toys of clay and different idols of gods and goddesses have been the tradition of our culture. But it is now regrettable that in recent times, especially in the last decade potters have been in distress. Because of these unavoidable factors like clay, lack of capital, unsatisfactory selling of clay pots, lack of fuel wood for burning raw pots, their plight is in peril.

Earthenware and fashionable things of clay are being rapidly supplanted by aluminum, plastic, steel and other alternative materials. Even toys for children are being made with wood and cloth. Besides, so cold prestigious people never tend to buy earthenware thinking their image and status. But it is admitted everywhere that cooking pot of clay is more conducive to health than pot of silver or other materials. Cooking rice of clay-pots help to cure gastric problem. And pitchers keep water cool in hot days. Another cause for not selling clayware is its brittleness. Inspite of being more cheaper than other aluminum or plastic made pots, clay-pots are not being sold available. Thus potters have to survive with a negligible earning.

A potter named Paras Chandra Pal told with rage, after liberation war many potter families had left the country away. The reason behind their leaving home allegedly are precarious future of pottery, oppression by neighbours as communal violence, political molestation & feeling of dire insecurity. He also informed that to bring money as a loan for capital from banks, they have to pay bribe to bank officials. NGOs often help them by providing loan with low interest. Potters are also concerned at the rising price of fuel wood and clay. Above all, their hard labour to pottery is not undeniable at all. Many of them grudgingly rush to adopt another occupations leveing pottery gradually.

Hence, government and connoisseur of pottery both should come forward to alleviate their poverty and evaluate their artistical work precisely. The commoner can also play a role merely considering the question of precisely. The commoner can also play a role merely considering the question of preservation of our Banglee tradition.

People live in the beautiful houses here made of clay. Roofs of them are made of burned clay called "tali" which protect them very well from the rain water.

Cattles are raised by almost every family

They are used for plowing lands, getting fresh milk, and in cow carts which are used to transport goods, specially rice, wheat, potato and other things produced on the fields to bring home.

Most Bangladeshi people, poor or rich, used to rely on clay pottery products, whether painted or painted, for their household ornament.

But as lifestyles changed along with social and economic advancement, the demand for clay pottery products plunged in Bangladesh as low-cost and metal-based substitutes became more in vogue.

The situation over the last few years, however, has gradually reversed since the potters started to make very attractive designs on pottery items and the quality and aesthetic values of their works have greatly improved.

"Clay pottery items for households were extremely close to extinction. Now we have brought changes to the design of the pottery items. So, people are showing interest in buying colorful potteries," Mofizul Islam, a seller, said.

He said customers' interest for clay pottery items and showpieces is growing day by day.

"I like clay potteries very much now that there are new designs to choose from," said Taslima Tandra, a young lady customer.

Tandra said that she prefers colorfully-designed clay potteries than ceramic items because the former appear to be more natural.

Customers and sellers have agreed that there is now a revival of the art of clay pottery-making. In fact, in Dhaka alone there are now hundreds of pottery shops producing a

variety of clay- based household items.

The Bangladeshi government has supported the efforts of Bangladeshi potters to regain its past glory.

As a sign of their interest, thousands of people have already visited the exhibition of pottery products organized by Bangladesh Folk Art and Crafts Foundation in Sonargaon. This annual month- long fair opened last Jan. 21.

Not just local residents but foreign tourists as well have flocked to the different stalls at the exhibition venue to buy ornaments made of clay.

Sonargaon, some 27 km east of capital Dhaka, used to be the capital of the Muslim Sultanate of Bengal from the 13th century until Mughal Emperor Akbar conquered the region.

The exhibition features a range of works by traditional potters from different parts of the country. An estimated half a million people from about six hundred villages in the country are directly engaged in pottery-making.

Works of both well-known and emerging talents are on display at the exhibition-- from household items like wall panels, tiles, functional tableware, light shades, crockery, fountains, garden decorations and other items.

In some stalls, potters display their talent in pottery-making to the delight of the customers.

In front of his would-be customers, Sushanta Kumar Pal, a 55- year-old clay artist, hand-painted his pottery item with diligence.

Pal, born in a potter's family in Rajshahi, some 261 kilometer northwest of Bangladesh capital, took up the ancestral profession at childhood.

"Once upon a time in Rajshahi, pottery items were very popular. About four to five thousand families were engaged in this profession. But now it is near to extinction over there," Pal said.

Pal, however, said that he will never abandon clay pottery making since it is their family heritage.

Although Pal never went to school for formal art education, his artworks highlight an array of styles, techniques, colors and textures.

Pal is now optimistic about the revival of this folk art as he called on the Bangladeshi government to help promote the industry by giving incentives to potters all over the country.

There are three features of the Bangladeshi landscape that assault the visitor as definers of culture: water, boats, and clay. The steady and dependable reaches of the rivers, estuaries, and the Indian ocean yearly yield to monsoon floods that sculpt the landmass by dumping a nutrient-rich silt, which supports the rice economy with three annual crops. With an agility born of long experience, the local populations not only adapt to the challenges of water, but make it the basis for their economy. The rich clays that provides the malleable structure to this water-logged land are pressed into analagous use, providing the renewable building materials for the village, the fundament of the country. With virtually no natural rock in this flood plain, it is no surprise that Bangladesh has always had an extensive community of artisans who specialize in this plastic medium. They not only master the clay to build complex shelter-ranging from the easily erected and maintained mud hut, with its distinctive Bengal roof, to enduring brick structures-but utiliarian objects, such as pots, food storage bowls, and other containers. It is a truism that wherever one finds an abundance of plastic material, art will follow its functional forms, and the potters of Bangladesh bear a long and rich history of decorative art-terracotta panels, glazed tiles, decorative pottery, toys, and images of divinity, gods, goddesses, and heroes, from the abstracted signifier to the icon steeped in realism. It is through this medium of clay@the pots and their potters-that Henry Glassie seeks to

26

introduce his reader to the larger culture of Bangladesh. It is a good choice because the potter's work touches everyone, but until now it has remained an unwritten chapter in the documentation of Bangladeshi cultural forms.

Glassie writes the book in an easy conversational style that quietly constructs a very personal image of contemporary society, each chapter alternating between descriptions of the places where the potters he interviews have traditionally and still now congregate-Dhaka, Kagajipara, Kakran, Shimulia, Rayer Bazar, and Shakharibazar-and the lives and experiences of the potters themselves. It is, in that form, a modified case-study ethnography. Many of these descriptions capture the rambling search of the author as he comes into contact with the sensory-rich experience of Bangladesh's physical and cultural landscape. Through this first-person discovery the reader becomes aware of the author's newness to Bengali culture (something that will mirror the average reader's position), with the effect of carrying the reader vicariously through some of the heady process of initiation experienced by anyone who steps into Bengal. …

KUMAR (potter) also known in Bangla as *kumbhakar*, is a traditional occupational group engaged in clay modelling and making earthenwares and various household items and toys from clay. Kumar is a caste name, which indicates that pottery as a profession was almost exclusively in the hands of Hindus in the past. The innumerable domestic wares prepared by kumars include *kalshi* (household water vessel), *handi* (cooking pot), *jala* (big water jar), *shara/dhakna* (pot covers), *shanki* (dish), *sharai* (jug), plates, cups, *badna* (water pot) and *dhupdani* (vessel for scented sulphur). Clay made toys and clay fruits like palm, banana, jackfruit or mango, are popular sale items in traditional Bangladeshi fairs and festivals. – BANGLAPEDIA

Continuously circulating human life is not that easy to explicate through literature. Life goes on in its own way. Motionless life will be mingled up with the clay. Motivation in life enhances the hope to live and breathe for future. Otherwise our life is nothing but a mist of suffocation. The hands working with the clay are the hands also struggling with the nasty rude world. But life goes on! Nothing can stop it. The hands now playing with the clay are actually playing with the eternal lives. Life goes on, life is endless! Life is circulating continuously!

In our everyday life we use many of the household items which are made by these potters. But we never take the time to appreciate their hard work or even to think how they are making these things day after day for us. They make most of these pots with their bare hands. They also take the help of a revolving wheel, which is made of wood or metal, to make these pots into different shapes and designs. The potter throws the kneaded clay into the center of the wheel rounding it off, and then spins the wheel. As the whirling gathers momentum, the potter begins to shape the clay. When it is over he severs the shaped clay from the rest. **Pottery is a skilled activity whereby different materials such as earth, porcelien and stones are used. Clay potters of Bangladesh, mostly of the hindu origin, called "Kumars" have been engaged in the practice of making pottery for a long, long time. Throughout the ages, these Kumars have been making pottery-ware manually using their hands mostly and with the aid of manually rotated wheel. However, during recent times, this age-old practice of manual pottery has been vanishing. The Kumars lead a hard life, often marred by poverty and need, as making a living out of pottery is a difficult one.**

Until recently, I never had the good fortune of actually seeing any potters or kumars in live action. I have seen plenty of television video footage and photographs of potters in action, and I had always wanted to photographs these talented people in action and thankfully, in a recent photography safari to the Shathkhira region of Jessore, Bangladesh, I had the good fortune of photographing them. Unfortunately, when I visited the potters' village, I was unable to witness them using the spinning wheel on which pottery is shaped as the Kumars don't use the wheel during the Bangladeshi Boishaki Season due to religious reasons.

Members of the potters' community, known in Bangla as kumars (formally known as kumbhokars) who usually bear the title Pal, are generally found in large settlements in mrit pollis (potters' villages) across the country.

The great ceramic tradition of Bangladesh unfolds in the context of geology. From the world's tallest mountains, mighty rivers roll to the sea. Their silt has built the world's widest delta. The earth of the delta is heaped into mounds that hold the villages above the flood. It is planted to rice so that people might eat. It is shaped and baked into vessels so that water can be carried, food can be cooked, and people can get though another long day.

There are six hundred and eighty villages dedicated to pottery-making in Bangladesh, nearly half a million people who use clay to make art because clay is what there is. They dig and mix two kinds of clay -- one white and sandy, one black and sticky -- treading and kneading them together to make a smooth new substance for creation.

The potters are predominantly Hindus. They bear the same surname -- Pal -- indicating their membership in the craft-craft-caste of the workers in clay. Most of the Pals make utilitarian ware. Women use *paras*, men turn the great *chak*. With their different techniques, women and men make identical *kalshis* and they collaborate in making *patils* -- smooth useful vessels that are slipped for brightness, fired to ruddy buff or silvery black, and then sent to market. During commercial exchange, Hindu products go to Muslim consumers, unifying society in the honorable ethic of utility.

Symmetrical, smooth, and bright, the potter's vessel is at once useful and beautiful. Then utility declines and beauty rises when the pot is painted. The *sakher bari*, painted by Hindu women in Rajshahi, is famed in Bangladesh, and at the Mirpur Mazar, the tomb of Hazrat Shah Ali Baghdadi, I watched Nur Mohammad paint *kalshis* that were crafted in the Hindu villages. Born in Mirpur in 1962, Nur Mohammad is one of the seventeen decorators who create souvenirs for pilgrims to the tomb. Working quickly to a geometric scheme, color by color, Nur Mohammad coats working pots with floral designs that evoke the promised garden of the Holy Koran.

Useful vessels dominate the potter's labor, but some potters specialise in images. Chittaranjan Pal was born in Bikrampur, a major source for the small earthenware statues that are sold in the markets of Dhaka and Chittagong. His father and grandfather made clay images, and Chittaranjan Pal now lives in Rayer Bazar, where, together with his wife, Mongoli Rani Pal, he molds, fires, and paints swarms of small statues.

Decorative images in clay subdivide into four main classes: animals, village life, heroes, and religious icons. The animals -- the birds and fowl, the horses and cows, the lions and tigers -- are most abundant. Toys for children to use in play, items to enliven the decor of the home, they can be read as merely decorative, though subtly they can carry deep messages. Is the handsome striped tiger a beast of the jungle or an emblem of the nation? Is the bird only colorful, or is it the symbol of the soul conventional in Sufi verse? Is the lion only imposing, or is it the *vahana* of Durga, the prime Hindu deity of Bengal?

Among the products for sale in the markets for pottery along the streets of Dhaka, enormous, vigorous horses prance out of the crowd. They were assembled of molded components, finished and fired in the workshops of Falan Chandra Pal and Joy Pal, of Maran Chandra Pal, Babu Lal Pal, and Narayan Chandra Pal, in the village of Khamarpara, Shimulia. The horse is a noble, decorative presence, suitable to the middle-class home, and it is a symbol of the mortal body that carries the soul through life as the horse carries the rider over the land. Conspicuously handsome and alive, the horse is a sign of vanity.

Maran Chand Paul is the great master of animal imagery in modern Bangladesh. He has taken traditional forms from the country markets and worked to give them a new refinement. Maran Chand Paul calls himself Mritraj, the King of Clay, and he is today the most famed among the potters of Rayer Bazar, where he was born in 1946. Like his father and grandfather before him, Maran Chand Paul was raised in the potteries, but his life took a turn when Zainul Abedin, the most renowned modern artist of Bangladesh, led his students on a sketching ramble through Rayer Bazar. Like the Swedish, artists who summered in the country, painting the people and the landscape, and whose fears for the death of the picturesque led to efforts at preservation and eventually, through the construction of Skansen, to the worldwide movement for open-air museums, Zainul Abedin and his students became interested in the people and in the preservation of their art. A collector of folk art who envisioned a museum for Bangladesh on the Skansen model, Zainul Abedin had served since 1949 as the principal of the Art College, now part of Dhaka University. He believed that Rayer Bazar's old pottery and tradition could survive only if it opened to include, as it has, Japanese techniques of shaping and English methods of glazing. Zainul Abedin brought Maran Chand Paul to the Art College. There Maran Chand Paul's education continued, and there he works as an instructor, using the knowledge he gained in his familial workshop to help his students realise their innovative plans.

With his colleagues in Rayer Bazar, Mohammad Ali and Subash Chandra Pal, Maran Chand Paul is a leader among the modernizing potters of Bangladesh. He is the master of a workshop that supplies toys for export, and he is an artist, widely recognised for two creations. The first, a part horse, was inspired by an earthenware horse from Bankura, West Bengal, which C.M. Murshid asked him to replicate when he was Pakistan's ambassador to China. C.M. Murshed chose a peculiarly Bengali form to use as an ambassadorial gift. It was a statement of cultural commitment in the ambit of political tensions between East Pakistan -- the Bangladesh of the future -- and West Pakistan. As it was for C.M. Murshid, the horse is a symbol of Bengal for Maran Chand Paul. It has proved to be such a successful commodity that he has created horses in many sizes, merging ideas from West Bengal and Bangladesh in the smaller versions, and he used the Bankura horse as a model in creating an earthenware elephant, the second of his signal forms. His elephant is symmetrically shaped in a mold, fastidiously finished by hand, and enriched with applied ornamental detail.

For their creators and consumers, earthenware birds and animals are depictions of the wonders wrought by God. They evoke the beauty and power of nature. The transformation of nature into culture is the topic of a second class of decorative image. It embodies affection for the human world, for the very land shaped and shared by Muslims and Hindus. The potter's form is a sculpture in relief: the single "wall plate" made for hanging , or the suite of "terracotta" tiles. The subject is rural Bangladesh: the river winding past forested banks, the houses of bamboo and thatch or corrugated iron, the spread of the agricultural landscape, and the common labor of women and men.

Subash Chandra Pal in Rayer Bazar and Nepal Chandra Pal in the village of Kakran have filled many commissions for terracotta portrayals of the lush landscape of Bangladesh. In the village of Kagajipara, the brothers Santosh Chandra Pal and Govinda Chandra Pal sculpt exquisite depictions of the land. The boats on the river, the houses in the village, the potters at work, and the Bauls in song -- are all closely observed and meticulously executed.

Amulya Chandra Pal, their neighbour in Kagajipara, is admirable for his bold style. He crafts images of animals. He sculpts pictures from the land, symbolising the nation through rural scenery and labor. And he symbolises the nation through portraits of its heroes. The heroes that the potters select for sculptural presence are the culture's great poets, its masters of the language.

Bangladesh is a place of beautiful views, and it is a place famed for the quantity and quality of its poetry. Amulya Chandra Pal is a poet as well as a potter. He was born about 1937 into an old pottery family, but his career did not truly begin, he says, until he married at the age of twenty and received encouragement from his wife's grandmother, Ashtrasukhi Pal. Amulya Chandra Pal has created many unusual images, including a portrait of Michael Madhusudan Dutt, the nineteenth-century Bengali poet who reconfigured myth into drama and epic, becoming one of those -- like his Finnish contemporary Elias Lönnrot -- who would form the national tradition into coherent verse before, through action, it could be shaped into political independence. But, like the other potters, Amulya Chandra Pal most often portrays the poets who make a pair in modern Bangladesh, as Shakespeare and Milton, rendered by the potters of Staffordshire, made a pair in Victorian England. One is Rabindranath Tagore, who won the Nobel Prize in 1913 and introduced

Bengali literature to the world. The other is Kazi Nazrul Islam, called the rebel poet, in whose veins, Mahatma Gandhi said, freedom flowed. One a Hindu, the other a Muslim, these are the men of our century who most beautifully shaped the language shared by all in Bengal.

Animals, village life, and heroic poets make three of the classes of decorative image. Fourth is the religious icon. For his Hindu customers, Amulya Chandra Pal sculpts and fires images of Saraswati, of Ganesh, of Radha and Krishna. For his Muslim customers, he creates depictions of mosques.

Molded and fired images of mosques and especially of the Hindu deities are made commonly by potters in Bangladesh. The religious topic of the fourth class of decorative image leads us toward the pinnacle of the potter's art in Bangladesh: the murti, the unfired image of the deity that is used in worship.

Both women and men make utilitarian vessels. Only men make murtis. Their work begins with a prayer, a mantra of praise that brings a direct revelation of the divine to the mind. Then as the artist of realism works to realise an image that registers on the retina, the maker of murtis works to realise an image that God placed before the mind's eye. As the artists describe it, God -- called Bhagaban, called Allah -- is omnipresent and without form. God appears in the mind in the form of a particular deity. The deities each have their spheres, their missions. In working to advance a mission of God on earth, the artist forms an armature of sticks, binds it with rice straw, and then covers the straw with clay, working to set into the world the image that appeared in his mind. His work is driven by a desire for beauty.

In realising the aesthetic of the murti, the artist pitches every motion toward unnatural perfection. Nature's implicit orders are extracted, refined geometrically, and materialised in a singing symmetry of form. Nature's roughness is erased in dampened layers of fine clay. Smooth and idealised -- exhibiting the qualities found generally in what is called folk art -- the murti is not a failed attempt to picture a woman out of the world. This is a goddess, perfect in beauty and not of this world.

Haripada Pal's work is an aesthetic act and an act of devotion. In the clay, he says, there is the seed of creation, the presence of God that springs to life with prayer. In his body, there is a drop of God, the soul that enables all action. As he slowly massages the clay, urging it towards perfect form, the power in his body surges through his fingers to unite with the seed of creation in the clay. This fusion of divine power is sustained by the dampness that abides in the clay. The potter of India burns the clay to purify it, but Haripada Pal says that fire kills the power that lives in the moist interior of the murti. The statue's interior is for power. Its exterior is for beauty.

When it is dry enough, the dark clay is covered with a coat of thickened white paint that seals the surface and prepares it for the application of luminous color. Bright paint completes the aesthetic programme, and on the day set in scripture, the murti is installed for worship. The murti is not God. It is a receptacle for God's power in the form of a particular deity: Durga or Kali, Saraswati or Lakshmi, Ganesh or Krishna or Vishnu. As the potter Amulya Chandra Pal puts it, Hindus do not pray to the murti, but through it, just as Muslims do not pray to the Mihrab in the mosque, but through it to God.

The murti is a prayer, a device of communication. As the artist called the deity into the mind with a prayer, now the Brahmin calls the deity into the clay with a prayer. The deity descends. The crowd presses forward, straining to take darshan. Their eyes, meet the eyes of the murti. The water in the body connects with the water in the statue. The soul in the body connects with the deity in the image. Communication becomes possible. Haripada Pal calls the murti a mediator. It opens a channel between the worlds. Requests are made, gifts are given and received, and then the puja ends. The drums cease, the incense smokes no more. The deity is gone, and the statue is empty, a pretty shell, bereft of power. It is carried to the riverside in a carnivalesque procession, and, to the ululation of the women, immersed. Unfired clay melts back into the water from which it came, becoming the silt out of which the kalshis and the murtis of the future will be made. The cycles continue. The rivers go on flowing.

Haripada Pal says that some people find it strange that he would work for a month to create a statue that will exist for only one day. But once the deity has left it, he says, the murti is of no more value than the body after the soul has flown. Some bury the body, others burn it, and the murti must be sacrificed in running water. His livelihood and his devotion require repeated creation. During creation, he unites with God, and he responds honorably to God's gift of talent by striving to make every murti as beautiful as it can be. Haripada Pal is a success in the world, respected in his community as a great artist, and his hope is that his effort will so please God that he will be released from the cycles of reincarnation and launched into a state of perpetual bliss.

Shaped of mud, then sacrificed in the river, the murti repeats the cycles of birth and rebirth. Its impermanence is a perplex to the historian of art. The potter's greatest creations do not last beyond the period of the ritual. Only the ethnographer on the spot gets to see and record them. But the potters also create for the temples. The temple to Kali on Shankharibazar contains a monumental image of the goddess, sculpted by Haripada Pal and ornamented by the goldsmiths of Tanti Bazar. Painted clay murtis are found in the temples, and it is fortunate for art historians that the potters use their modeling skills to shape figures that are cast in metal. The ancient bronze and the modern work in brass or copper is a tangible shadow of the murti in clay.

Sankar Dhar works in a shop a few blocks from Haripada Pal's in Old Dhaka. He was born into a family of carpenters in 1957. His father worked for the temples, framing the lion seats, the ornate wooden stages that shelter the deities. Sankar Dhar continues his father's line in wood, but early in his life he began to work in clay. He was not taught, he says; he learned by experimentation. Had he received a degree from an educational institution, he believes, his talent might have brought him wealth, but he comforts himself by observing that the great poets Rabindranath Tagore and Kazi Nazrul Islam were, like him, self-educated, and in his art he fulfills his inner passion. At work in the clay, Sankar Dhar makes murtis for pujas, and, like Haripada Pal, he sculpts human figures to be used in the displays in shop windows. Haripada Pal once made a stunningly realistic portrait of Bangabandhu Sheikh Mujibur Rahman, who would have been the Washington or Ataturk of Bangladesh if his bright life had not been cut short by brutal assassins. Sankar Dhar also excels at portraiture, and he has made clay models of the deities for sand-casting in brass. His gleaming image of Durga, cast in the scared blend of eight metals, stands now in the sanctuary of the Dhakeswari Mandir in Dhaka.

Sankar Dhar's creations in wood and metal carry us away from work in the clay, but the potter's art has provided a pattern, derived from Bangladeshi tradition, that will help organize our thinking as we shift to exploration of other media. Utility is not belittled; it is stressed. The potter's common vessels are useful, and so is the potter's greatest work. The murti is used by people to connect with God so that they might receive the blessings that make life on the earth tolerable. All the works are marked by symmetry in form and by smooth, bright surfaces. Then they divided by function into the plain tool, the ornamented tool, the decorative image, and the sacred image. And the third of those classes, the decorative image, subdivides into animals, village life, heroes, and religious icons.

Terracotta, often mistaken as clay tiles, is actually baked earth (pora maati). There are specific themes and subjects based on which terracotta art is produced. A variety of items is made of terracotta – plaques, wall tiles, lamps, pitchers, flower vase, pottery, coin bank, candle-stands, dolls, and more. The designs of these burnt clay products reflect folk tales, picture of everyday life, artistic symbols denoting peace, love and understanding. In Bangladesh, some of the most contemporary and prominent pieces of terracotta art are dominant on plaques and murals portraying our Liberation War–or rural Bangladesh depicting many birds, fish and animals and agricultural activities.

Terracotta artists say, although this art is one of the oldest in the sub-continent and was spread out across the whole of this country, is practised in only a few prominent localities today – Dhamrai, Shimulia, Kagojipara, Kakran in Savar; Baufol in Barisal; Shariatpur; Mirzapore in Tangail; Rayer Bazar in Dhaka (almost extinct) and a few other remaining places in the country where pottery villages exist.

This art was practised in Bengal from the earliest through early medieval to medieval times and even persisted in Hindu monuments till the mid-nineteenth century. It was found in the earliest civilisations like Harappa and Mohenjodaro. The Vishnupur (17th century) temple of (West Bengal) and Kantaji's temple (18th century) of Dinajpur are the finest examples of this form in this region. Ancient terracotta murals are found in their finest forms in Paharpur, Mahasthangarh and Maynamati dating back from the 6th century to as far back as 2000 A.D. Some plaques depicting ladies with exotic hairdo found in Mahasthangarh throw enough light on the affluence and elegance of the people inhabiting the region. Most recent specimen of the art can be found in Pabna, Jessore, Faridpur, Rajshahi, Barisal and some other places in Bangladesh.

The clay which is used are derived from Barisal and Savar called Doash and Etel maati referred to as primary and secondary clay, amalgamated in a mixture, often called common clay. These clays can withstand high temperature when inside the furnace and make the best quality of products.

Other potters use clay from different parts of the country. Terracotta artisans say that a type of white clay is collected from Netrokona, reddish clay from Baufol Barisal and pinkish clay which comes from Tangail and these different soils produce different types of products and add their own texture and colour to the items after baking.

The whole process can be explained like this: these different types of clay used for making terracotta items are collected from different regional locations. The assorted materials are mixed proportionately and dried under the hot sun so as to allow any sort of moisture that may be present in it to evaporate. Then, the mixture of this wet clay is filtered through a fine sieve to remove pebbles, dirt or even rice grains.

After the products are given shape with the hands, they are baked in improvised kilns in very high temperature ranging from 850 to 1375 degree Fahrenheit and gradually burnt for almost 24 hours. The products cool down inside the oven when the fire is put off and it is maintained strictly so that no air can enter the furnace. It takes almost 40 to 45 days for a batch of terracotta products to be made properly following the rules starting from clay amalgamation up to the final cooling.

Today, there is a strong terracotta revival movement going on in the country. Md Azharul Islam Sheikh Chanchal, assistant professor of the ceramics department, faculty of fine arts, Dhaka University, said, " Some organisations have proper production units, design studios and are helping this revival process. Prominent among them are - Vertical, Burn Clay, Harappa, Pora Mon, Clay Image and a few more in the city."

"There are some celebrated terracotta artists and one of the most renowned is Moron Chad Pal sir, an ex-teacher of our fine arts faculty and a pottery artist by family tradition. He is considered one of the biggest assets of the country who produces tepa putul (clay dolls). Other mentionable pottery artists are Mohammad Ali from Rayer Bazar, Shuhash Pal, Bisheshor Pal from Baufol Barisal, Robi Pal, Ruhi Pal. They are absolutely classic artisans capturing the essence of the heritage efficiently", Chanchal added.

These organisations are mostly artists' collective with the objective of reviving people's interest and also to revitalise the traditional pottery and terracotta industry of Bangladesh.

One such organization is Harappa Ltd and this correspondent spoke to their production director and in-house designer Rudro Naser. He said, "We are now hiring the potters from different locations of the country, bringing them to our organisation in Dhaka and training them not only on making the pottery of their interest but other burnt clay products as well and be creative with whatever they do. These kumars are usually a frustrated lot because their works in their specific areas are not selling any more, for example clay-made household utensils. Our motto is to save the potters by introducing a few modern techniques, mechanical wheels and train them to think of designs out-of-the-box".

One such frustrated potter is Bishwanath Chandra Pal from Tangail, who is now working in Dhaka. "I came to Dhaka in 2006. Clay pottery was our ancestral business but during the early 2000s, we realised that this trade will not fetch us food any more and so tried to divert to other trades. But when I was offered a spot to come to the capital city and work in the same trade to earn a living, I immediately took the chance."

"The biggest difference between working at home and here is that, in Tangail, we always stuck to one form of terracotta product, for example coin banks. Year after year, we made the same thing and nothing else. But here, with training of moulding techniques and using a more modern wheel and different types of clay, we make a range of products – wall panels, tiles, tableware, fountains, lamp shades, crockery and other decoration pieces", Bishwanath added.

'Shokher Hari' is one of the oldest traditional clay products in the country. And this is the story of another Sushanta Pal from Rajshahi, who courageously decided to preserve the arts of terracotta in his own way. His grandfather Banyeshwar Pal was a famous exponent of pottery art and received provincial awards during the Pakistan period.

But today, due to deep frustrations over the decline on the use of clay potteries and to run his family, this craftsman transformed this art form from pottery to art paper and canvas, believing to preserve this traditional art and as well earn a living.

Sushanta said: If clay pottery art is dying then indeed it is a grave news for us because this traditional art captures the essence of rural Bengal and sports the images of mainly folk tales, celebrations, the different plants, birds, fishes, other animals and of course, humans at their best.

"Size and names of painted potteries differ based on regions, the style of painting and the design motifs – Mongol Hari, Jagoron Hari, Aiburo Hari, Phul Hari and the extravagant Shokher Hari. And the name Shokher Hari was coined in Rajshahi region which cultivates exceptional value of art and grandeur", Sushanta added.

And Sushanta Pal craftily used to depict this art on earthen bowls and pots and recently realised that these clay wares started losing its force to non-artsy contemporary bowls. "I started to transform this art on thick art paper primarily and canvas to preserve our roots, our culture and glorious, colourful identity and to earn a living in this world of mechanised civilisation. Yes of course the other material pots has great value of longevity but are they emotional pieces of articles?" asked the frustrated Pal.

The great artisan Moron Chad Pal, thinks terracotta having mother earth as its raw material has a unique power of soothing. These products can be displayed in any surroundings. They signify peace in terms of colour, textures and have the unique ability to control the temperature of the atmosphere. Its cool composition brings a certain tranquillity and organic feel to the person using it.

"The significance of terracotta is therefore not restricted within the walls of temples and mosques or even the clay dolls and potteries but they send this urgent message of going back to our roots", an emotional Moron Chad added.

There were 'Kumarpara' or 'Palpara' - of potters in almost all the villages of rural Bangladesh since ancient times. An integral part of Bangladesh's culture for hundreds of years, pottery has now become popular interior decoration materials across the country, specially in the cities.

Once used in day-to-day lives of the people of the country for cooking and storing water or religious purposes, pottery items are now used as show pieces and adorn posh homes.

Though many shops in posh shopping centres of Dhaka sell pottery items, the footpath near the Bangladesh Shishu Academy in the city is the favourite hub of those who wish to buy items for decorating their homes. A good number of nursery plant shops alongside help increase the sale of shops of pottery and terracotta showpieces.

Affluent people of the city, university students and tourists mainly visit the shops and buy terracotta items of their choice. "Pottery items and terracotta showpieces are very popular among the city people. Buyers generally opt for this place as this is the biggest market of pottery and terracotta items,' said a shopkeeper.

New terracotta showpieces including wall hanger, sculpture, flower vase and ashtray are also available in different shops. Earthen ornament sets, flower vase, wall hangings, earthen 'banks' for collecting coins, candle stands, clay mugs and bells, animal figures, replicas of brides and grooms can be found near the Shishu Academy and also in the posh shops.

The price of pottery varies according to their design and size. A small item can fetch at least Tk 150 if it is of good quality.

Lately there has been a trend of producing replicas of famous paintings of Shilpacharya Zainul Abedin and Patua Qamrul Hasan.

Pottery items are already being exported. It is hoped that if it receives the attention of the government and easy loans from the banks, this ancient craft would become popular household decoration in many parts of the world.

Potters in Bangladesh are facing a hard time in Bangladesh these days. The hereditary craftsmen have to deal with multifarious problems: a slump in sale, high price of clay, lack of capital in addition to inadequate patronage from the concerned authorities.

Traders say that earthenware and clay products are rapidly being replaced by aluminium, plastic, steel and other alternative materials. Even toys for children are being made with wood and cloth. Besides, many of the urbanites think it is beneath them to buy earthenware. However, those who know point out those clay utensils are far more hygienic and environment friendly than those made of silver or other metals.

Mintu Paul, a hard-pressed potter, said that after the Liberation War (1971), many artisans left the country. Confronted with dipping sales, many in the potter communities are now considering alternative professions.

Statistics bear out Paul's contention. Over the last 30 years the number of potter families have decreased alarmingly. Many have migrated to India or taken to alternative professions such as shop-keeping, rickshaw pulling and rowing boats.

Or take the case of Phul Mohammad. Around 15 years ago Phul Mohammad migrated to Dinajpur town from the Khansama village for better prospects. However, his skills as a potter have not been appreciated; today his offsprings are in search of more lucrative professions.

Even as demand for high quality earthenware registers an upswing, former potters such as Bhuttu Paul of Biral upazila, Dinajpur says that his community lacks the know-how to fashion such products and market them to the urban buyers.

Bangladeshi potters have always laid stress on the basic form and texture of his articles. Harmonious color blending, the perfect all-over effect of design with shade and tone, mark his unity of purpose.

As for the types of wares, pottery comprises true distinctive types of wares. The first type, earthenware, has been made following virtually the same techniques since ancient time; only in the modern era has mass production brought changes in materials and methods. Earthenware is basically composed of clay- often blended clays - and baked hard, the degree of hardness depending on the intensity of the heat.

After the inventions of glazing, earthenwares were coated with glaze to render them waterproof; sometimes glaze was applied decoratively. It was found that, when fired at great heat, the clay body became non-porous. This second type of pottery, called stoneware, came to be preferred for domestic use.Many kumars manufacture bricks and tiles, along with earthenware of all shapes and sizes, and idols and toys. The factories of kumars well repay a visit. Beneath the same thatched roof are the kiln, storehouse and dwelling house, while a free space in front of the door is used as a place to prepare the clay. In the rainy season boats laden with earthenware from these places travel to neighbouring districts through the rivers.

The traditional pottery industry all over the district is on the verge of extinction, throwing hundreds of potters in utter helplessness. Meanwhile, a large number of potters have already quit their ancestral profession being unable to cope with the problems engulfing the age-old industry.

Besides, the demand for pottery products has decreased in the local market due to easy availability of stainless steel products in the markets to a considerable extent. With the passage of time it is apprehended that the artisan community of the district will have no other alternative but to adopt other manual jobs for their survival